BRITAIN'S INNER CITIES

Paul Lawless is a Principal Lecturer in Urban Planning at Sheffield City Polytechnic. He has taught extensively on urban studies courses in Sheffield and London and previously worked in several planning offices in London. He has published a number of books on inner city policy and local economic development.

BRITAIN'S INNER CITIES

SECOND EDITION

Paul Lawless
Sheffield City Polytechnic

P·C·P
Paul Chapman
Publishing Ltd

To Anna and Ruth

First published 1989
by Paul Chapman Publishing Ltd
144 Liverpool Road
London N1 1LA

British Library Cataloguing in Publication Data

Lawless, Paul, *1948–*
    Britain's inner cities. – 2nd ed.
    1. Great Britain. Cities. Inner areas.
    Social conditions
    I. Title
    941'.009'732

    ISBN 1–85396–039–X

Typeset by Inforum Typesetting, Portsmouth
Printed and bound by Athenaeum Press, Newcastle upon Tyne

B C D E F G H 9 8 7 6 5 4 3

Contents

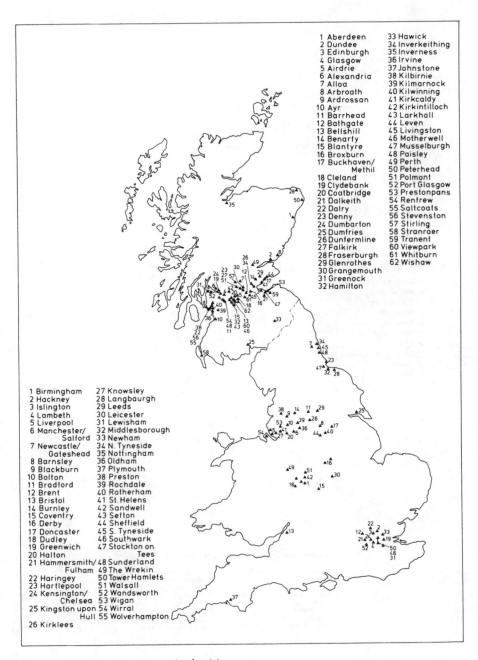

1 Aberdeen
2 Dundee
3 Edinburgh
4 Glasgow
5 Airdrie
6 Alexandria
7 Alloa
8 Arbroath
9 Ardrossan
10 Ayr
11 Barrhead
12 Bathgate
13 Bellshill
14 Benarty
15 Blantyre
16 Broxburn
17 Buckhaven/
   Methil
18 Cleland
19 Clydebank
20 Coatbridge
21 Dalkeith
22 Dalry
23 Denny
24 Dumbarton
25 Dumfries
26 Dunfermline
27 Falkirk
28 Fraserburgh
29 Glenrothes
30 Grangemouth
31 Greenock
32 Hamilton

33 Hawick
34 Inverkeithing
35 Inverness
36 Irvine
37 Johnstone
38 Kilbirnie
39 Kilmarnock
40 Kilwinning
41 Kirkcaldy
42 Kirkintilloch
43 Larkhall
44 Leven
45 Livingston
46 Motherwell
47 Musselburgh
48 Paisley
49 Perth
50 Peterhead
51 Polmont
52 Port Glasgow
53 Prestonpans
54 Renfrew
55 Saltcoats
56 Stevenston
57 Stirling
58 Stranraer
59 Tranent
60 Viewpark
61 Whitburn
62 Wishaw

1 Birmingham
2 Hackney
3 Islington
4 Lambeth
5 Liverpool
6 Manchester/
   Salford
7 Newcastle/
   Gateshead
8 Barnsley
9 Blackburn
10 Bolton
11 Bradford
12 Brent
13 Bristol
14 Burnley
15 Coventry
16 Derby
17 Doncaster
18 Dudley
19 Greenwich
20 Halton
21 Hammersmith/
   Fulham
22 Haringey
23 Hartlepool
24 Kensington/
   Chelsea
25 Kingston upon
   Hull
26 Kirklees

27 Knowsley
28 Langbaurgh
29 Leeds
30 Leicester
31 Lewisham
32 Middlesborough
33 Newham
34 N. Tyneside
35 Nottingham
36 Oldham
37 Plymouth
38 Preston
39 Rochdale
40 Rotherham
41 St. Helens
42 Sandwell
43 Sefton
44 Sheffield
45 S. Tyneside
46 Southwark
47 Stockton on
   Tees
48 Sunderland
49 The Wrekin
50 Tower Hamlets
51 Walsall
52 Wandsworth
53 Wigan
54 Wirral
55 Wolverhampton

Figure 1   Urban Programme Authorities

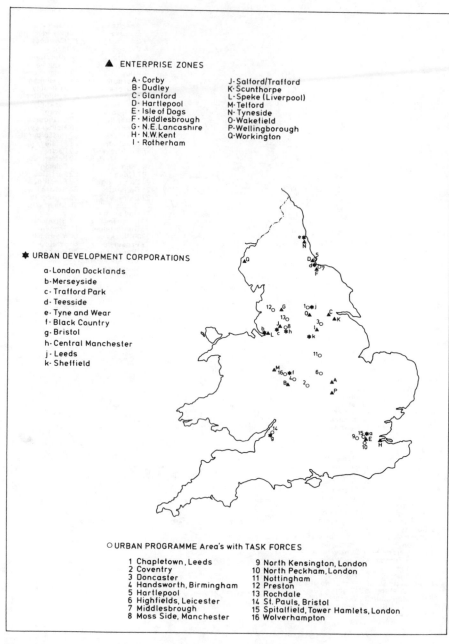

▲ ENTERPRISE ZONES

A· Corby
B· Dudley
C· Glanford
D· Hartlepool
E· Isle of Dogs
F· Middlesbrough
G· N.E. Lancashire
H· N.W. Kent
I · Rotherham

J· Salford/Trafford
K· Scunthorpe
L· Speke (Liverpool)
M· Telford
N· Tyneside
O· Wakefield
P· Wellingborough
Q· Workington

✳ URBAN DEVELOPMENT CORPORATIONS

a· London Docklands
b· Merseyside
c· Trafford Park
d· Teesside
e· Tyne and Wear
f· Black Country
g· Bristol
h· Central Manchester
j· Leeds
k· Sheffield

O URBAN PROGRAMME Area's with TASK FORCES

1 Chapletown, Leeds
2 Coventry
3 Doncaster
4 Handsworth, Birmingham
5 Hartlepool
6 Highfields, Leicester
7 Middlesbrough
8 Moss Side, Manchester

9 North Kensington, London
10 North Peckham, London
11 Nottingham
12 Preston
13 Rochdale
14 St. Pauls, Bristol
15 Spitalfield, Tower Hamlets, London
16 Wolverhampton

Figure 2   Urban Development Corporations and enterprise zones in England

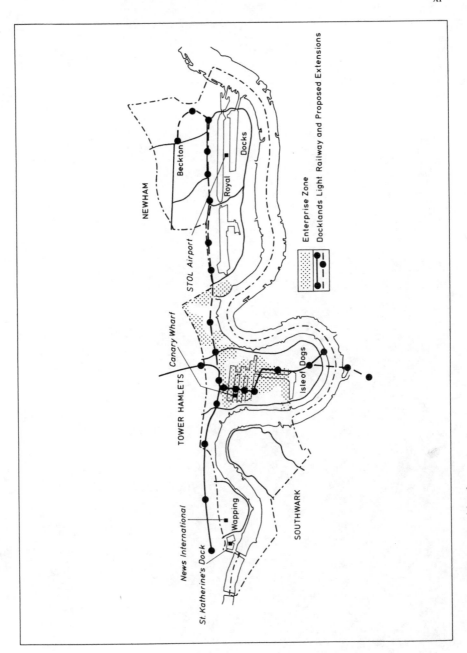

Figure 3 London's Docklands

Preface

This second edition is long overdue. The first edition was published in 1981 but written in 1979 and 1980. It was out of date before it saw the light of day. This will always be a problem with books that address contemporary policy development. However, inner-city policy has changed so much that assessments based on the political events of the late 1970s are largely irrelevant to the urban debate of the late 1980s. This book is, therefore, not so much a revised edition as a totally rewritten edition: not one sentence survives from the original book. It would also be nice to think that the policy development so apparent in inner-city intervention in the ten years or so since the first edition of this book had actually meant something to deprived people living in the major conurbations. On the whole, unfortunately, this is not true.

The book adopts a threefold structure. In Part I, inner-urban policy is located in its historical context and in contemporary social, demographic, economic and political parameters. Part II, the largest section, examines inner-urban policy after 1977. Chapter 3 evaluates the tentative early steps taken by Labour government between 1977 and 1979. The next three chapters assess approaches adopted by Conservative governments since 1979. Although the resulting classification has flaws, I have tried to deal with developments since 1979 thematically rather than chronologically. Chapter 4 looks at co-ordinating projects; Chapter 5 deals with programmes based on liberalization and enterprise; and Chapter 6 is concerned with initiatives intended to improve urban development. Chapter 7 outlines some local-government innovations in the field of inner-urban policy and Chapter 8 is dedicated to non-governmental agencies. Finally, in Part III, some fundamental themes are explored. Chapter 9 presents an overview and critique of inner-city intervention as a whole; Chapter 10 looks at some explanatory devices that might be used to understand the development and evolution of inner-city policy; and Chapter 11 outlines an agenda for reform.

Paul Lawless
Sheffield City Polytechnic,
1988

PART I

Inner-Urban Policy: the Context

1

The Origins of British
Inner-Urban Policy

The 1977 white paper, *Policy for the Inner Cities* (HMSO, 1977), can be seen in retrospect to have heralded an era of permanent British inner-urban policy. Since 1977, both Labour and Conservative governments have initiated a series of urban policies and programmes that are the main focus of this book. However, this period of more substantial inner-city policy is clearly rooted in a number of urban experiments undertaken in the late 1960s to the mid-1970s. Many of these projects are of little more than historic interest, but some are of greater significance. Some continued, if in modified form, after 1977. Others were to explore and to highlight the scale of decline in many parts of urban Britain. A few of these projects were to prove influential in the evaluation of existing, and in the definition of new, models of urban deprivation – conclusions that were to inform inner-urban policy after 1977.

From 1966 to 1977, about a dozen separate experiments were developed, primarily by the Department of the Environment and the Home Office, and details of these projects, their structure, administration and conclusions have been explored elsewhere (Edwards and Batley, 1978; Higgins, 1978; Lawless, 1979). In most cases, these experiments have limited contemporary relevance, but three issues are worth exploring: Why was an urban programme ever introduced? What major experiments were implemented? And what were their key conclusions?

Why were Urban Experiments Initiated?

A number of factors can be seen that encouraged both Wilson's 1960s Labour government and Heath's 1970–4 Conservative government to introduce urban experiments (McKay and Cox, 1979). Interestingly, many of these initial factors remain relevant to the continuing debate about the urban problem and its resolution. Unemployment was rising in the cities – London, for example, lost half a million of its 4.5 million jobs between 1961 and 1975 and most major provincial cities lost at least 30 per cent of their manufacturing jobs in the five years after 1971 (DoE, 1975).

While job loss was undoubtedly one factor that encouraged central governments to experiment in the urban field, other impulses were also important. Independent surveys (Abel-Smith and Townsend, 1965) and official reports (Plowden, 1967) confirmed that twenty years of the Welfare State had not eradicated poverty. To do so, imaginative new programmes for the disadvantaged would need to be sustained in the cities by central government. This had certainly happened in the USA, whose urban programmes were evaluated and – to some extent – promoted by key British civil servants. In particular, US poverty programmes introduced by Presidents Kennedy and Johnson were more intensive and controversial than anything attempted in Britain, but some of the assumptions and policies contained in the American experience were assimilated by British policy-makers.

Racial issues cannot be ignored in this discussion. From about 75,000 in 1951, the non-white population in Britain rose to almost 600,000 in 1966. Controls on immigration had been introduced by a Conservative government in 1962, and were tightened by the succeeding Labour administration. Despite these controls, however, certain national and local politicians, notably Enoch Powell in a notorious speech in April 1968, continued to express anti-immigration and anti-immigrant sentiments. These sentiments have been seen by some observers (Edwards and Batley, 1978) as a primary factor in encouraging Wilson's government to instigate an urban programme in 1968 – legislation contained within the Local Government Act 1966 had, in fact, already allowed central government to provide grants for staff costs incurred in dealing with some of the transitional problems caused by New Commonwealth immigrants. The urban programme announced by Wilson in 1968 was more intensive. It was not solely designed for areas with large numbers of immigrants, and it allocated substantially larger resources in the 1968–77 period than those available under the 1966 Act.

One other issue that ought to be mentioned in this context is the localized nature of many of the urban experiments. A number of reports in the 1960s argued that particular problems were prevalent in certain areas and that policy initiatives, therefore, should be directed towards specific localities. The Milner Holland Report, for instance, suggested that the problems of stress and over- crowding in parts of London merited the establishment of areas of special control (Milner Holland, 1965). Plowden (1967) argued that compensatory educational resources ought to be directed towards deprived areas in cities to counteract their adverse environmental and domestic circumstances. Similarly, Seebohm (1968) advocated co-ordinated social planning in areas of special need. The assumptions behind these and other publications were influential in the urban debate. From the early developments of an urban dimension in the 1960s, through to the policy innovation of the late 1980s, the assumption has often been that particularly-grave urban problems were concentrated in definable urban localities, and that policies designed to deal with these problems should be defined in terms of locale. We shall return to the validity of this approach shortly.

Other influences encouraged the formation of urban experiments in the 1960s and 1970s and, it can be argued, these helped to foster the later, more-permanent urban policy. The creation of new urban projects was politically appealing – experiments could be instigated at limited cost but with undeniable political prestige. Governments were seen to be both caring and innovative. Experimentation allowed politicians (notably Secretaries of State for the Environment) to gain personal political acclaim though the initiation of what were often low-cost projects. It allowed central government to influence policies that were traditionally the domain of local authorities. Urban experiments also offered the prospect of co-ordinating the myriad of policies and agencies that in some way or other influenced the governing of cities. This last factor should not encourage the idea, however, that rational forces of reform drove forward the era of urban experimentation. To those observing the earliest development of the urban programme, the entire initiative seemed unwarranted and largely the result of pragmatic, *ad hoc* political necessity (Edwards and Bately, 1978) – an epitaph similarly fitting to urban policy after 1977. Indeed, many of the themes discussed above remain pertinent to inner-city intervention throughout the 1970s and 1980s, as is discussed in Chapter 10.

The Major Urban Experiments

A number of the urban experiments instigated in the late 1960s and early 1970s were designed to explore specific problems of city life, and to develop policies pertinent to them. Some were concerned with the quality of life in urban areas; others examined the ability of area management to alleviate poverty and to improve urban administration. While these projects helped to illuminate the scale of the urban problem and the applicability of policy innovation in certain areas, they had but little lasting impact. The same could be said, rather more surprisingly, of the educational-priority area (EPA) concept that emerged from the Plowden Report (1967). EPAs were established, and some additional resources were allocated to them, but the entire debate about urban education has never been central to urban policy. Perhaps more should have been done in this area. However, as the most notable research programme on EPAs pointed out, many deprived children did not go to schools in EPAs and, although enhanced educational investment was important, it had to be seen as part of inter-related policy initiatives that also included such things as housing and employment (Halsey, 1972). But if the EPA concept had only a limited effect on the evolving pattern of urban intervention, four other experiments proved significant in either helping to structure post-1977 policy or in raising issues that have remained relevant to the inner-city question. Urban aid or, as it is often called, the Urban Programme, should be mentioned because of its administrative and financial importance during the period of urban experimentation. Projects designed to improve the co-ordination of urban governance – notably the Comprehensive Community Programmes (CCPs) – continue to be relevant. The Inner Area Studies and the Community Development Projects were influential in re-orientating assumptions towards urban poverty in the mid-1970s.

The original Urban Programme ran from 1969 to 1977, when it was substantially restructured as outlined in Chapter 3. During this eight-year period, Home Office circulars requested bids from local government and the voluntary sector to fund a range of community, social and educational projects. The resources allocated typically amounted to about £20 million per annum, normally never sufficient to cope with demand, despite the fact that central-government funding was 75 per cent of total costs. There were benefits from the programme. Thousands of individual projects were supported, including many in the voluntary sector that would not otherwise have occurred. There were, however, important defects as well. Evidence suggested that authorities used the Urban Programme to fund projects that would have been implemented in any case (Holman, 1971). Also, there was no systematic review of what the programme was designed to achieve, how it would achieve it or how its progress would be assessed (Edwards and Batley, 1978). This failure to link action and research severely undermined the programme's effectiveness and impact. In financial terms it was the largest of the urban projects initiated in the 1960s and early 1970s; its longer-term consequences for the urban debate were minimal.

The CCPs were, in a sense, the mirror image of the Urban Programme: their contemporary significance was limited but their underlying premise – the search for co-ordination of urban agencies – has proved a perennial theme. To the incoming Labour government of 1974, it appeared that too many disparate urban projects were being developed with little in the way of overall co-ordination (Spencer, 1980). Other observers had come to much the same conclusion. Shelter, for instance, working in Liverpool, had earlier argued (1972) that the multiplicity of problems affecting parts of that city necessitated a 'total' approach in which relevant agencies should implement an agreed programme of action. The Inner Area Studies (see p. 7) were also partly designed to test the viability of co-ordinating strategies intended to moderate urban disadvantage. But of all the urban experiments, the CCPs were perhaps the most structured attempt to test ideas of co-ordination.

The details of the CCPs – of which only two were implemented in England (at Gateshead and Bradford) – are discussed elsewhere (Spencer, 1981). There were basically to be three parts to the programme: part 1 would outline the nature of deprivation in an authority; part 2 would analyse policy in relation to the needs of deprived areas and specific client groups; and part 3 would outline projects to be implemented in the succeeding financial year by central and local government. In the event, the CCPs were largely marginalized within the two authorities concerned: they proved unable to carve out substantial executive powers and were frequently misunderstood by officers and councillors. CCPs themselves were abandoned in 1980, but the assumption remained that urban deprivation could be moderated through the amalgamation of the activities of relevant agencies.

Undoubtedly, the most important of the urban experiments were the Inner Area Studies and the Community Development Projects (CDPs). They began to publish their more significant findings towards the middle of the 1970s, and their collective conclusions on the causes of – and responses to – urban deprivation were profoundly influential on the early development of permanent inner-city policy after 1977. Twelve CDPs were initiated by the Home Office between 1969 and 1972.

Unlike the Urban Programme (which scattered resources widely), the projects examined community issues in smaller areas of between 10,000 and 40,000 people. Nor were they all in urban areas. They consisted of community action thrusts that would be evaluated by structured research programmes. They varied enormously in their attitudes, in their responses to parent local authorities, in their prescriptive policies towards moderate disadvantage, and so on (NCDP, 1974; NCDP, 1975). The rigorous evaluation of models of urban deprivation undertaken by some projects – notably at Coventry, Saltley (Birmingham), Newcastle upon Tyne and Tynemouth – had an impact on the wider debates that related to deprivation and its causes.

So, also, did the Inner Area Studies (IASs). These, however, were more cohesive than the CDPs. There were three of them – in Liverpool, Birmingham and Stockwell (Lambeth). They were undertaken by consultants commissioned in 1972 by Peter Walker, Secretary of State for the Environment. They were given a wide brief and in their conclusions they presented a generally-reformist strategy (DoE, Birmingham IAS, 1977; DoE, Lambeth IAS, 1977; DoE, Liverpool IAS, 1977). Existing administrative and financial arrangements would suffice to re-invigorate the areas under consideration, providing political support was forthcoming and, importantly, providing more resources were allocated to the area. Money was needed for housing renewal, retraining, educational investment, infrastructural support and to create and to sustain local jobs. The Lambeth IAS proved slightly maverick in its conclusions, arguing strongly for a programme of balanced dispersal that would allow, simultaneously, the more deprived to move out of inner London and to encourage some of the better-off to remain there – a strategy others were to subscribe to a decade later (Buck, Gordon, and Young, 1987). But the real significance of the IASs, together with the CDPs, was that they helped to transform attitudes towards urban deprivation: assumptions held by central-government ministers in the mid-to-late 1960s were invalidated in the 1970s. The IASs and CDPs were vitally important in this process.

The Urban Experiments and Models of Deprivation

The lasting impact of the urban experiments was that they were prepared to investigate and, in many cases, to reject attitudes towards urban poverty held by central governments in the late 1960s and early 1970s. They helped to shift the debates from over-simple models of deprivation towards more profound and far-reaching attitudes regarding the urban problem, attitudes that were to guide inner-city policy for many years. We can best explore this process by analysing three models of deprivation: the culture of poverty thesis; co-ordination; and structuralist positions.

The Culture of Poverty Thesis

The culture of poverty thesis was developed in America by such writers as Banfield (1970). This theory assumes that anti-social behaviour is transmitted from generation to generation in families concentrated in certain parts of cities. These assumptions were assimilated by British ministers. For instance, in 1968, Callaghan – the Home Secretary – argued that 'there remain areas of severe social deprivation in a number of our cities and towns – often scattered in relatively small pockets' (1968a) and that urban aid was 'intended to arrest, in so far as it is possible by financial means, and reverse the downward spiral which affects so many. . . . There is a deadly quagmire of need and apathy' (1968b). In the early 1970s, these sentiments were reiterated by, among others, Sir Keith Joseph, Secretary of State for Social Services. He argued that a culture of deprivation characterized by early marriage, early child-rearing, poor educational attainment, vandalism and petty crime could be identified in parts of Britain (Joseph, 1972). Inadequate parents produced inadequate children. This cycle of poverty had to be overcome by improved preparation for parenthood, by better educational facilities, by health visiting, and so on. Not surprisingly, in this climate of opinion some of the early urban experiments – notably the Community Development Projects – were evaluated according to the degree to which they helped moderate social ills such as desertion, divorce and child abuse (Home Office, 1970).

To any government investigating questions of urban deprivation, there were obvious advantages in pursuing the idea that anti-social behaviour engendered by community or individual *malaise* in certain inner-city areas was the root cause of disadvantage. As Sinfield pointed out at the time, 'an all out attack on a number of specific areas is much more administratively attractive – certainly cheaper and potentially quicker than the careful re-examination of the basic fabric of society' (1973, p. 134). Certainly, those advocating the culture of poverty thesis tended to emphasize the apparent weaknesses of individuals and inner-urban communities. Wider questions about, say, the role of the disadvantaged in the economy, or issues of wealth and power, were ignored. However this over-simple, if elegant, notion was substantially weakened, if not destroyed, as a credible explanation for deprivation in the mid-1970s.

There were a number of reasons for this re-orientation in official thinking on urban poverty (Lawless, 1979; 1986), and some of the key developments should be mentioned here. The CDPs and, to a lesser extent, the IASs, were significant in this respect. The Coventry CDP, one of the earliest to publish its final conclusions, was instrumental in evaluating the culture of poverty thesis (Coventry CDP,

1975). Hillfields, the area selected for analysis, did not appear to the team to be different from other parts of inner Coventry, and there was little to suggest that local inhabitants were in some way inadequate. Poverty certainly existed, but it was clearly associated with – if not caused by – variations in employment capacity in the local vehicle-manufacturing industry. These conclusions were echoed by many of the CDPs and IASs that were collectively unable to identify pockets or areas of poverty within cities where especially-deprived and inadequate communities had survived. This finding was confirmed by research emanating from the Department of the Environment. The *Census Indicators of Urban Deprivation* and its associated publications (see, for example, Holtermann, 1975) examined the spatial manifestations of disadvantage as revealed by the 1971 Census. This research particularly highlighted the widespread nature of deprivation and the very limited spatial coincidence between different indicators of poverty. Bad housing, for example, tended to be concentrated in older, often privately-rented accommodation. However, the highest rates of unemployment were frequently encountered on the newer, public-sector estates. Hence, the 1971 Census did not highlight convenient small areas to which additional resources could be directed.

The idea that poverty was a much more widespread and complicated phenomenon than the culture of poverty thesis would suggest was reinforced in 1976 by the publication of Rutter and Madge's *Cycles of Disadvantage*. This work was commissioned by Sir Keith Joseph in the early 1970s, and it was intended that it should highlight the extent of intergenerational continuities in a range of aspects of deprivation. However, far from concluding that intergenerational continuities characterised poverty, the research team indicated that, certainly in the economic field, there was a surprising degree of mobility both upwards and downwards between generations.

The collective conclusions of the CDPs, the IASs, the *Census Indicators of Urban Deprivation* and *Cycles of Disadvantage* were important in undermining culture of poverty ideas. It was now difficult to continue to blame individuals and communities for their undeniable poverty. Other forces had to be found, and these are discussed on p. 11. However, two final points should be mentioned. First, it is apparent that intergenerational continuities in poverty do exist, particularly in certain urban areas. This raises broader questions of why such poverty is initiated and sustained. Secondly, although the culture of poverty thesis received little credence in the urban debate after the mid-1970s, very similar thinking continued to permeate policies adopted by Conservative governments in the 1980s. For example, the restructuring of the welfare benefits system in 1988 was based on the idea that a deserving poor should be supported to the detriment of an undeserving rump. There are clear parallels here with the culture of poverty's assumption that the urban poor had only themselves to blame.

Co-ordination

Throughout the late 1960s and early 1970s, issues of management, efficiency and co-ordination featured prominently in social-welfare debates (Holman, 1978). Technical inefficiencies, a lack of integrated service delivery and the failure of key services were often seen as increasing poverty.

These types of assumptions were absorbed into the urban question. As we saw earlier, some urban experiments (notably, but not exclusively, the CDPs) were based on the idea that urban deprivation might be substantially relieved through the better management of existing resources. In the early 1970s this thinking fitted neatly into the development of improved managerial techniques in local government. The Bains Report (1972) argued that local authorities should attempt to devise overall corporate strategies that might be updated frequently as a result of close monitoring. Instead of line departments reporting to largely independent committees, an overall strategy would be created to embrace all local-authority activities. Similarly, Bains and other commentators (Royal Town Planning Institute, 1976) promoted the idea that area management would improve the delivery of services to the more deprived. This managerial approach took many forms (Mason *et al.*, 1977). However, the approach was based on the decentralization of some local-government activities and personnel into 'mini-town halls'. Powers and financial resources granted to the area teams, their structure, administration and briefs varied widely. But the underlying concept was clear: in common with corporate planning, and the activities of some of the key urban experiments, deprivation might be alleviated by the improved management and co-ordination of services.

There may be advantages in this approach. The proliferation of urban agencies may suggest the need for a co-ordinated and unified strategy towards the cities. There is also, no doubt, scope within local government and elsewhere for greater efficiency. It may make little sense for either different branches of local or central government to pursue different or contradictory policies – this happens within local authorities and, at the national level, the Department of the Environment and the Department of Trade and Industry have not always had the same views on the policy, direction and even desirability of inner-urban intervention.

On the other hand, many of those who have evaluated attempts at better management within local government or attempts to instil improved co-ordination through urban experimentation, hold an altogether more jaundiced view (Cullingworth, 1974; Benington, 1975; Coventry CDP, 1975; Holman, 1978). Their criticisms tend to be either technical or conceptual in nature. In the case of techniques, it is apparent that improving the delivery of services and co-ordinating agency programmes is difficult. Different agencies have contrasting views on policy direction and tend to operate within different time-scales, according to different political incentives. It is also evident that an over-emphasis on management and efficiency can be seen as diversionary, directing attention away from such vital issues as resources. Instead, it tends to stimulate the growth of committees, reports, deadlines, flow charts, and so on, which may improve the internal management of organizations but have little impact on the deprived. It could be argued that improved co-ordination is not what the cities require. A variety of approaches and strategies need to be adopted. No single approach has been so successful that all others should be abandoned. Nevertheless, the search for co-ordination and management have continued to play a prominent role in urban policy, as later chapters show.

Social, Economic and Political Structures

Throughout the urban experiments of 1968–73, central-government ministers (as has been outlined above) were – at least ostensibly – of the view that deprivation was best explained by cultural arguments. Perhaps the most lasting influence of the CDPs and IASs was that they proposed different and more complex explanations. Instead of focusing on the apparent inadequacies within individuals, families and communities, the urban experiments tended to place the disadvantaged within the prevailing social, political and economic parameters. While not particularly original, this focus certainly helped to re-orientate inner-urban intervention away from an over concern with individuals towards a consideration of the forces that contributed to urban disadvantage.

An important theme was the determination of the experiments – notably the CDPs – to examine the effect of welfare benefits and allied services on poverty (NCDP, 1974; NCDP, 1975). The 'culture of poverty' arguments tended to assume that deprived individuals and communities should be encouraged to raise themselves out of poverty. It rapidly became apparent that for many poor people this was a somewhat naïve supposition. Economic parameters imposed severe constraints on many traditional working-class communities. In any case, substantial groups in society – such as the elderly and single-parent families – had no contact with the labour market: they depended on social benefits. To those exploring the mechanics of social benefits, the system seemed inadequate and complex (Bradshaw, Taylor-Gooby and Lees, 1976). Hence, the urban experiments argued that for substantial groups in society, the main cause of deprivation lay in the operations of the welfare-support system.

These conclusions did not have the same influence on the evolution of the urban programme as, say, economically-orientated findings. There may be a number of reasons for this. For example, there was nothing specifically 'urban' about defects in social security. If it was a problem, it had national implications. The Labour government of the 1970s, and Conservative governments elected after 1979, were not going to be easily persuaded that inner-urban policy required a substantial integration and enhancement of welfare benefits. If anything, national governments in power after 1976, when the International Monetary Fund imposed reductions in public expenditure, were in the business of trying to reduce social-security payments. Urban policy after 1977 was not concerned with aspects of social welfare and, increasingly, not especially with urban poverty.

This is undoubtedly a central weakness in inner-city policy. By 1983, over 7 million people in the UK were either on Supplementary Benefit or on incomes within 20 per cent of that level. About 60 per cent of these were pensioners, single people of working age or one-parent families (CSO, 1987). Not all of these people can be described as poor, but many are. For many in the cities and elsewhere, their best hopes of improved material circumstances lie not with the labour market but with welfare benefits. In the late 1980s that is not a particularly welcome position in which to be.

In many respects, political alienation, also identified by the urban experiments as a constraint on many of those living in the cities, has been similarly neglected by inner-urban policy. The CDPs and IASs emphasized the importance of incorporating individuals, communities and organizations (such as trade unions) into the urban political debate. The National Community Development Project, for instance, argued in a radical vein that dis-investment in traditional manufacturing industry would need to be countered by positive trade-union activity (NCDP, 1977).

More reformist sentiments were to emerge from the IASs. The Birmingham study wanted a more sensitive response to local needs through greater devolution of political responsibility (DoE, Birmingham IAS, 1977). Whatever the merits of these sentiments, it is apparent that inner-urban intervention has been concerned only minimally with attempts to provide disadvantaged groups with greater political responsibility or power. The urban experiments may have identified political irrelevance as an alienating constraint on many of those within the cities. Nevertheless, little within inner-city policy has concerned itself even peripherally with this issue, despite the fact that to official observers (Scarman, 1981) and independent commentators (Rex, 1982), the interweaving influences of political alienation and economic disadvantage were key factors in explaining the urban disturbances of 1981.

If social and political constraints operating on those in the cities received only limited attention in the urban policy effected after 1977, the third major structural force identified by the urban experiments – economic decline – was a major feature of permanent inner-city intervention, the urban experiments themselves largely agreeing that this was the most acute constraint affecting the cities. To the Lambeth IAS, it seemed that job opportunities and the residents' ability to earn a living were fundamental (DoE, Lambeth IAS, 1977). These factors had a wide impact. Declining individual, household and community income worsened poverty, increased social ills and reduced spending on a whole range of both public and private goods. Wealth and income were, in effect, the primary generators of inequality. Any effective urban policy must address the whole jobs, wealth and income problem. Health, housing, social services and education were also important, but the central issue behind urban decay was economic decline.

The implications of this analysis were profound. In the late 1960s and early 1970s, central-government ministers could use the culture of poverty argument, but not everyone believed the over-simple messages inherent in it. Academics expressed considerable doubts (Townsend, 1974). It also may be presumed that central governments were eager to uphold the cultural arguments – whatever their intellectual limitations – because of their undeniable advantages to any national administration: the problem wasn't widespread; it required limited resources; and the major challenge was on the need for individuals and communities to raise themselves out of poverty and not to rely on State intervention.

Structuralist positions, however – stressing the primacy of economic issues and the effect of wider problems on the cities and those living in them – contained more profound implications for government. The problem was not spatially limited. By the late 1970s, declining employment and economic output were characteristic of many parts of Britain, and it would be expensive to control, let alone reverse, this decline. The blame for fewer job opportunities could hardly be placed on individuals, communities or even entire cities – broader forces of economic rationalization were at work.

Nevertheless, despite the worrying implications in structuralist diagnoses, the 1977 white paper appeared to accept these. It asserted 'that the decline in the economic fortunes of the inner areas often lies at the heart of the problem' (HMSO, 1977, para. 7). In succeeding years, the concept that urban decline would require the implementation of policies designed to boost economic output has been central to inner-urban policy. However, this programme of urban intervention has been influenced by a range of political, economic, financial and social forces that operate on and within the cities. These are addressed in Chapter 2.

2
The Urban Context

Chapter 1 outlines some key historical developments in the evolution of urban policy. However, since 1977 various other forces have been at work in the cities and, consequently, have affected the efficacy and relevance of inner-urban policy. Indeed, it could be argued that the collective influence of the processes discussed in this chapter have made a greater impact on the cities and those living within them than has inner-city policy. These forces can be grouped into four major categories: political considerations, socio-demographic factors, economic change and financial constraint. Each is discussed below.

The Political Environment

The fundamental issue in this context is the breakdown in the mid-1970s of the post-war political settlement, the subject of considerable analysis (see, for example, George and Wilding, 1984; Raban, 1986). In brief, programmes advocated by Beveridge and Keynes were accepted and implemented by national governments from 1945 until the mid-1970s. These programmes were founded on a comprehensive Welfare State system complemented by the demand management of an expanding mixed economy. This 'middle-ground' approach received bi-partisan support until the mid-1970s, when sharply-divided views began to emerge. Collectivist principles were still seen (by at least some on the left) to provide a coherent and viable strategy that would both sustain economic growth and support the Welfare State. This approach was very much at the heart of Labour's 1983 election manifesto. Alternatively, from the mid-1970s, a new libertarian strand of thinking came to dominate the Conservative Party. State intervention, collectivism and political consensus, which had characterized politics after 1945, were seen as significant causes of economic decline and social dependency. A new ideology based on individualism, a diminution in public activity and intervention, free enterprise and selective welfarism prevailed within a Conservative Party that was about to gain electoral success in 1979 and throughout the 1980s.

This apparent breakdown in the post-war consensus has been attributed to a number of inter-related processes. Some have argued that international economic slumps, notably that after the oil-price rise in the early 1970s, imposed unacceptable constraints on the ability of weaker national economies – such as Britain's – to support a comprehensive Welfare State (Mishra, 1984). Other commentators, such as Barnett (1986), place a greater emphasis on the interweaving influence of economic decline and political ineptitude that, even in the 1930s, prevented governments from facing up to the constraints imposed by the vested interests of major corporations and trade unions. Other observers place the collapse of the post-war settlement in a narrower ideological framework, arguing that a separate ideology of the Welfare State can be identified independently of capitalism and socialism (Pinker, 1979).

Whatever the ultimate causes of this breakdown, it is clear that – for many of those able to perceive a substantial division in post-1945 British political attitudes – important changes occurred in the 1970s. The first change occurred in 1976 when the International Monetary Fund imposed reductions on public expenditure, and the second with the election of the first Thatcher government in 1979.

Inevitably, this brief summary of a major debate surrounding change and continuity in British politics ignores a whole range of issues. There is no universal agreement, for example, that the mid-1970s represented a fundamental break in political attitudes – aspects of the 'new' capitalism coming from the Conservative Party in the mid-1970s can be traced back to Churchill's administration in the early 1950s (Raban, 1986). However, these and similar discussions are beyond the scope of this book. It seems reasonable to assume that there was a change in the dominant ideology within the Conservative Party in the mid-1970s. Clear differences in approach can be identified between the last two years of Heath's 1970–4 administration – with its strong *dirigiste* tendencies – and policies effected by Conservative governments elected since 1979. These later governments have advocated (if not always implemented) programmes designed to reduce collectivism, public-sector activity and expenditure, and apparent social dependency, whilst liberating enterprise and encouraging individualism.

The ideological changes within the dominant political party since 1979 have undoubtedly affected in various ways the cities and those living in them. A national economic policy designed to reduce public expenditure, to liberate enterprise and to eradicate inflation, combined with a determination to limit local-government spending and intervention, was extremely damaging for the cities and their economies in the early 1980s. In addition, other policies had an adverse effect on the major conurbations. This may not have been the explicit aim of the programmes outlined below, but the consequences for the inner cities were real enough.

One way in which changing State ideology has been felt in the cities since 1979 is the determination of central governments to reduce and even eradicate public-sector policies and organizations. The Housing Act 1988 is likely to reduce the stock of public-sector dwellings substantially. This is likely to prove particularly significant in those cities with the most council accommodation. There may be a case for greater flexibility in the housing market, but a reduced public sector and probable higher housing costs will do little to reduce the national homelessness figures of over 100,000.

It is also apparent that the cities have suffered as a result of the abolition in 1986 of the metropolitan counties and the Greater London Council. The government argued (HMSO, 1983) that these conurbation-wide bodies were unnecessary and irrelevant. Some were perhaps not unhappy at the abolition of the counties, for certain powers returned to the lower tiers of government. However, any rational analysis of urban administration must conclude that the lack of any elected conurbation-wide bodies – particularly for London, the West Midlands and Greater Manchester – is shortsighted. Metropolitan counties are needed to guide the strategic planning of the city regions, to organize intra-urban transport and to frame economic and demographic mobility within and between conurbations. Their abolition increases costs, encourages an over-proliferation of uneconomic developments, stifles the mobility of the poorest and generally inhibits rational urban planning.

The problems of encouraging orderly development in the older conurbations have been heightened not only by an inappropriate metropolitan administrative structure, but also by an inadequate and diminishing regional influence. The social and economic performance of a city inevitably reflects – although it is not exclusively determined by – the fortunes of the larger region in which it is located. Regional performance is influenced, if only at the margins, by regional policy. Since the 1930s, regional policy has attempted to direct, on the whole, manufacturing companies to the more depressed regions of the north and west of Britain. Between 1960 and 1976 regional policy possibly created nearly 300,000 jobs in the Assisted Areas (Moore, Rhodes and Tyler, 1977; Fothergill and Gudgin, 1982). The bulk of these jobs were not located in the inner cities – many went to green-field locations and new towns. Real costs were incurred in redirecting jobs away from more prosperous areas, and not all of the jobs that were relocated survived periods of economic slump. Nevertheless, in at least some areas regional policy undoubtedly moderated unemployment. However, after 1979 this success failed to keep regional policy intact, when it reflected all too clearly the hand of State intervention.

Early indications of Conservative attitudes to regional policy were given in 1983 (DTI, 1983). Regional policy was still seen as a significant factor in remedying imbalances in unemployment. In part, this could be achieved by widening regional policy to encompass aspects of service employment (not a traditional focus for regional support) and by linking approved projects with job creation. These changes, implemented in 1984, might in themselves have been seen as distinctly advantageous. However, total regional support was down-graded as both levels of support and areas granted regional status were reduced.

Despite the government's determination to modify regional expenditure in the mid-1980s, total regional aid – both selective assistance and the more important automatic development grant – amounted to almost £700 million in 1986–7. Perhaps partly as a result of this and perhaps partly as a result of the government's view that regional aid should be more immediately directed towards assisting business skills and enterprise within smaller companies, a further major review of regional policy was published in 1988 (DTI, 1988). Among the other revisions it proposed, the Department of Trade and Industry argued that the automatic grant for buildings and machinery that was available through regional development grant was to disappear. Selective assistance, where companies demonstrate that assistance is necessary for the implementation of specific projects, was to remain. The government argued that this was correct since many projects would have gone ahead irrespective of automatic grant. Not everyone was convinced. Labour MPs and some key Conservative figures, such as Michael Heseltine and Leon Brittan, argued that a removal of the automatic grant would worsen regional imbalances (Duffy, 1988). Obviously, companies prefer automatic grant, but that is not a justification from the government's point of view (Begg and McDowell, 1987).

This loss of automatic grant might prove particularly important when regions attempt to attract major international inward investments. The Nissan car plant near Sunderland is a notable example of this. In 1988 the production of a second model from the plant was supported by a £25 million grant, which helped to provide another 1,400 jobs. Selective assistance may not be able to attract as many developments. Many inward investments can be criticized for lacking research and development facilities, for having limited local independence, and so on. However, many projects – from America and Japan in particular – are destined for the European market. If UK incentives prove unattractive, these mobile projects will locate themselves elsewhere within the Community.

This brief discussion of regional policy may seem somewhat divorced from inner-urban problems, but it is, in fact, very relevant. It demonstrates how Conservative administrations in the 1980s have been determined to reduce public expenditure and to ensure that support is ultimately dependent on central-government approval – a constant theme in inner-urban policy, which later chapters explore. Debates relating to regional policy also highlight the way in which the economic problems of the older industrialized regions, which include most inner-cities, have not figured prominently in government thinking since 1979. Regional policy may not be a major strand of governmental intervention and in the short to medium term it is destined to decline in importance. But its evolution, combined with trends in housing and urban administration referred to earlier, gives an indication of the political constraints acting upon the cities in the 1980s.

Another factor to be felt in the cities is the increasing political marginalization of the older industrial regions of the UK. Elections in 1979, 1983 and especially

1987, showed starkly the political polarization in Britain. While southern England and surburban Britain voted Conservative, by 1987 the major cities (except Birmingham and London) returned Labour MPs almost unanimously. This is not surprising – Labour offered far more to those in the cities than did the Conservatives. Investment in housing, transport, education and social services would have risen with a Labour victory. However, despite an effective campaign, Labour secured just 32 per cent of the vote, leaving the Conservative government with an overal majority of 102.

There are a number of possible reasons for Labour's election failure (Curtice, 1987; Kellner, 1987). Conservative policies, such as the sale of council houses, proved popular with many working-class voters. In the South of England, where political power increasingly resides, the anti-Conservative vote was split between Labour and the Alliance. Perhaps Labour was also unable to appeal to many voters in an increasingly better-off, more obviously middle-class Britain.

Whatever the reasons for Labour's weak performance in the 1987 election, few working within urban government can regard this as anything but unfortunate. Labour's reformist programme to moderate poverty and to re-invigorate the older industrialized regions – whatever its ultimate success – at least offered more to the urban disadvantaged and to the cities generally. Towards the end of the 1980s, it is increasingly apparent that cities are likely to find themselves very much in the firing line, as a series of radical measures are implemented concerning education, social security and local taxation. These measures seem certain to remove local government from aspects of educational provision, to reduce massively social security payments to those in the inner-cities – for example, by more than 80 per cent in Liverpool and Glasgow (Brindle, 1988) – and to create financial and social hardship through the imposition of the poll tax. A further Conservative victory in the early 1990s might be avoided either by an electoral pact between Labour and the Democrats, or by increased support for Labour, particularly in southern England. Neither of these developments seems likely. Perhaps the most realistic possibility for a more reformist and supportive programme of urban support lies in a re-orientation of attitudes and policies within the Conservative Party. However, the radicalization of the party suggests this will not be forthcoming.

The continued political marginalization of the cities and their inhabitants raises the question of urban disturbances. The 1980s saw urban riots on a scale unprecedented in peacetime twentieth-century Britain. Major civil disturbances occurred in 1980 in Bristol, in 1981 throughout the UK and in 1985 at Broadwater Farm in Tottenham and again in Birmingham. In 1981 more than 4,000 people were arrested in serious incidents of public disorder(Home Office, 1982). Inevitably there has been considerable speculation on the causes and implications of urban disturbances. In the context of the apparent marginalization of the cities, there is a central reformist tradition (developed in both Britain and the USA) that suggests that riots need to be seen as political statements of dissatisfaction. This sentiment is particularly evident in the Kerner Commission's report that explored riots in American cities during the 1960s (Kerner *et al.*, 1968) and to some extent in the Scarman report that focused on the Brixton disorders in April 1981 (Scarman, 1981). These analyses examined both the immediate and the more fundamental causes for urban disturbances. In the Kerner report, police tactics and the role of the media were seen as especially significant. But the causes for the disturbances lay elsewhere.

These and other commentators basically concluded that riots reflected a disturbing degree of political alienation by a substantial minority of inner-urban inhabitants, particularly young, single men. Very often these are not the most deprived in society. In the UK, about half of those arrested in 1981 were employed. Most, despite popular belief, were white. Some urban riots in the 1980s had racial connotations – whites fought blacks and blacks have fought the police – but in general most disturbances involved urban youths, both black and white, in confrontation with the police. To Kerner and Scarman the one factor explaining this confrontation is the economic, physical and social environment in the older urban cores: housing and educational standards are deficient; jobs are often hard to come by and badly paid; orthodox political parties appear irrelevant; and public and private-sector services seem inadequate. Alienation and even violence become understandable, almost predictable responses.

Scarman's position in 1981 was well received by many political commentators. The report concluded that a reform of existing institutions and policies would alleviate substantially the poor conditions under which many were living in Brixton and, by implication, in other inner-urban localities. Hence, a predisposition towards violence would be moderated. Reform could take many guises, but there would have to be large investment in housing renewal, educational provision and job creation, while steps would need to be taken to reduce the real but, Scarman argued, not institutionalized, racism apparent within the Metropolitan Police. Reform would need to be deep rooted, but existing structures could be modified. Radical change was, therefore, not on the agenda – no doubt a conclusion very much to the satisfaction of the government.

However, there must be some reservations about Scarman's reformist programme. On one level it would be difficult to argue that the kinds of interventionist policies supported by the report were ever carried out in Brixton: such inner-city areas as this suffered from declining public-sector resources in the mid-1980s. At a deeper level, it is apparent that not all observers share Scarman's analysis of the problem or its associated policy response, and it is possible to identify stances both to the right and left of this central reformist tradition.

Many within the Conservative Party and a large segment of the popular Press considered that the urban disturbances were related to, if not caused by, immigration. Young black people were seen as engendering an anti-British culture that, because of its criminal overtones, inevitably lead to conflict with the forces of law and order. Criticisms of this approach have been documented by, for example, Lea and Young (1982): most of the rioters were not black; some riots occurred in areas where there were very few black people; most black people are British; a 'British' culture seems elusively difficult to define; and so on.

Other explanations for urban disturbances are based in class as well as race. In the case of class, commentators such as Friend and Metcalf (1981) locate the riots within broader economic processes. In the long post-war boom, labour shortages throughout Europe and North America necessitated the introduction of new groups into the job market, particularly women and immigrants. In Britain's case, the position of immigrants was different from that of most European countries as Commonwealth immigrants had certain civil rights. Even though immigration was controlled from the early 1960s, those immigrants and their families in the UK were entitled to remain. This was unlike the position of, say, the Turks in Germany. Immigrants to the UK fulfilled in many ways the same functions as white working-class people. A reserve army of labour was created that included many Afro-Caribbean people and people from the Indian sub-continent. These and their white peers were taken on in times of expansion. After 1976, however, the long contraction in economic performance, combined with processes such as technological change, led to a steadily-larger reserve army of unemployed labour. It was this process of economic polarization that fuelled the urban disturbances of the 1980s.

This review of the urban disturbances is important as it places the cities and the urban labour forces in broader economic trends. Much of this thinking has emanated from neo-Marxist debate, a paradigm that is reiterated in Chapter 10. However, this interpretation of the urban disturbances should be treated with some caution. Evidence from both the UK and America suggests that the most deprived tend to be less involved in riots than those slightly higher up the economic ladder. This type of observation suggests, perhaps, that the most accurate commentaries on the urban disturbances have emerged from those adopting positions somewhere between neo-Marxist interpretations and the central reformist tradition (Rex, 1982). There is clearly an economic impetus to riots: urban employment and the lowering of individual and community wealth and income increase the probability of urban riots.

But other forces need to be considered. Racial intolerance and racism are probably greater than Scarman assumed. There is evidence that institutionalized racism is apparent within the police (Policy Studies Institute, 1985; Broadwater Farm Inquiry, 1986). In addition, a whole range of aspects of consumption, housing, education, shopping, transport, training, and so on, remain insufficient for many people living in the cities. These deficiencies give rise to dissatisfaction in those who can see increasing inequalities in a society where suburbanites and those in the smaller towns and cities of the south of England have experienced large material gains through house-price increases and tax changes. These inequalities inevitably give rise to resentment. The types of minimal reforms advocated by the Scarman report will not suffice in moderating this scale of resentment. To bring the argument full circle, one frequently-neglected aspect of the urban problem is the political dimension. The urban disadvantaged have little in the way of an effective political outlet. Power resides elsewhere, in other places and in parties that do not need urban parliamentary seats. It is not surprising that political activity by working-class city dwellers has declined sharply in recent decades. Any serious attempt to resolve the urban problem cannot ignore questions of political participation within the cities and the urban political debate's impact on the national stage.

The Cities and Socio-Demographic Change

The major demographic change occurring in recent decades within the cities has been a decline in absolute population levels (Redfern, 1982; Champion, Coombes and Openshaw, 1983). There are inevitable statistical complexities in defining what 'urban' means in this context. However, there was certainly a considerable population loss from the cores of the larger, older provincial cities from 1971 to 1981. Some inner-London boroughs, such as Kensington and Chelsea, lost in a decade more than one-quarter of their 1971 population total. Glasgow, Manchester and Liverpool all lost more than 15 per cent and even free-standing cities such as Nottingham and Hull were losing around 5 per cent. To a certain extent, the declining population totals reflected falling birth-rates in cities, but the main factor here is population decentralization. The scale of this process and its considerable effect on the cities raises a number of pertinent issues. Why have so many left the older urban cores? What implications does this have for the cities? And what future trends might be anticipated?

Why have People Left the Cities?

A distinction can be made between planned and unplanned decentralization. Most households left the cities through unplanned decentralization, but some were moved as a result of specific planning policies. For example, some were moved to the edges of the older urban cores as a result of the construction of peripheral public-housing schemes in the 1960s and early 1970s. Cantril Farm in Knowsley, 8 miles from Liverpool, is one particularly notorious example. These developments often failed because of the limited scale of the local social and economic infrastructure. However, most who moved out of cities as a result of planned decentralization did so through the new-town programme.

The construction of 28 new towns after 1946, housing more than 1.1 million people by 1987 (Potter, 1987), must be seen as the major positive achievement of town and country planning in the UK. The debates concerning their development are well documented (for example, Aldridge, 1979) but in this exploration of the inner-urban problem three issues should be raised. First, although it was always assumed that many from the older urban cores would be rehoused in the new towns, early evidence (Heraud, 1966) shows that it tended to be the better-housed suburbanites who moved, few from the inner cities even knowing of the programme. Second, new towns offered at least some public housing for the urbanities who were unable to buy their way out of the cities. Thus, the abolition of the new-town development corporations in England in the late 1980s has strong implications for the mobility of the poorest in the cities: they will not find a great deal – if any – of non-owner-occupied housing beyond the conurbations. Third, only a small proportion of those leaving the cities moved to new towns, perhaps only 10 per cent. Hardly any left cities such as Leeds and Sheffield, around which no formal new towns were commissioned.

Most of those leaving the cities have done so through the commerical market and they have moved for a variety of reasons (Kennett and Hall, 1981): more freely-available, cheaper, owner-occupied housing might be found beyond the cities in environmentally-attractive locations; households are more mobile – car-ownership rates doubled between 1961 and 1981 and the electrification of some InterCity lines has encouraged a marked decentralization of people away from London to areas such as Peterborough, Stamford (Lincs.) and even Newark (Notts.); many move out of cities on retirement; and for the economically active in the south of England, movement out of London becomes ever more attractive as many commercial activities leave the capital. This process of corporate restructuring is discussed further later, but the overall effect in terms of demographic patterning is that many of those previously employed in London can now find equivalent work in the free-standing towns of the south. Since 1961, the expansion of towns and cities within a hundred or more miles of London has been enormous. In the 1970s, for example, places such as Newmarket (Suffolk) and Huntingdon (Cambs.) grew by almost 20 per cent. In the following decade, communities to the west and south of London, such as Basingstoke (Hants), Reading, Guildford and Slough, were to expand even faster. Demographic decentralization may well have preceded the movement of jobs out of the cities but, by the mid-1970s, a self-sustaining centrifugal force was clearly draining the cities of both people and jobs.

What are the Implications of Urban Decentralization?

The scale of urban decentralization after 1961 has important implications. For regions such as East Anglia, which grew in population by more than 10 per cent in the 1970s, there are problems associated with growth. An increasing population, particularly one scattered in an unplanned fashion throughout a region, increases substantially the cost of service provision. Infrastructural, educational, health and retailing costs rise. Some costs fall on the private sector, but most fall on overstretched public-sector organizations.

However, the implications of urban decentralization for the cities are of greater concern. In some respects it could be argued that diminishing populations need not necessarily be unwelcome in the cities: renewal can take place at lower densities and much needed environmental improvements can take place. On the other hand, if city-based populations continue to decline problems will ultimately occur because of under-utilized services and diminishing taxation bases. That stage, however, has perhaps not yet been reached. The real problem is not so much the scale of decentralization but the selective nature of demographic decentralization.

A large proportion of dwellings constructed beyond the cities since 1960 have been in the owner-occupied sector. Most of those leaving the cities have had to buy their way out if they wanted to live in environmentally more-attractive areas that, particularly in the south of England and – to some extent – increasingly elsewhere, were proving to be the most economically active. Public-sector housing provision, which might have allowed for greater mobility, declined in the 1970s and 1980s. In 1979, 32 per cent of dwellings in Great Britain were in the public sector. By 1984 that figure stood at 27 per cent (CSO, 1981; 1987). This decline was due to both the sale of council dwellings and to fewer starts, which – at perhaps 20,000 or so in the mid-1980s – represents about one-fifth the early 1970s total. Much of what was started was implemented not by suburban, predominantly Conservative councils, but by urban Labour ones. However, the public sector offers little possibility of enhanced mobility for the urban poor; the same is true of private rented accommodation that, by 1984, represented only 12 per cent of dwellings in England and Wales. An overwhelming proportion of private rented accommodation is to be found in the major cities, notably London and Liverpool. Many of those living in the cities are trapped: they cannot afford to buy their way out, particularly not to the more prosperous areas of the south and rented accommodation – which is in any case largely urban based – is declining nationally.

Not surprisingly, areas to the west of London were encountering severe shortages of skilled and even unskilled labour by the mid-1980s, a time of real and substantial unemployment in the older industrialized regions. Job availability in the south is of little value when inter-regional mobility is difficult. A lack of mobility on the part of many within the cities has inevitably meant that an increasing number of those remaining are in some way disadvantaged (Redfern, 1982; Office of Population Censuses and Surveys, 1984). The older conurbations' cores tend to accommodate larger proportions of the unemployed. They contain fewer households with cars, more mentally and physically disabled and more of those with limited educational qualifications. Additionally, the cities often contain much larger numbers of the elderly than national figures would suggest, and far fewer from the higher socio-economic categories. Liverpool in 1981, for example, had 13 per cent of households in SEGs I and II compared with a national figure of 23 per cent. Some cities housed larger proportions of households whose head was born in the New Commonwealth than national statistics would indicate. In 1981 approximately 4 per cent of the British population could be so classified, whereas in more than fifty largely-urban parliamentary constituencies, the comparable figure was 15 per cent or more. This population concentration may reflect a number of factors, for example, cultural considerations. However, job opportunities and the availability of rented accommodation may also encourage households originating in the New Commonwealth to locate themselves within the older conurbations. For those from Afro-Caribbean backgrounds, prevailing material circumstances are possibly worse, on average, than for those pertaining to white working-class urbanities.

This concentration of the more disadvantaged within the cities has fuelled debates about the combined issues of the extent of multiple deprivation in the cities and the relative severity of that deprivation within the various cities. It is not simply that a large number of multiply-deprived people live in the urban cores (Hall and Laurence, 1981); there are individuals and households where this applies, but highest unemployment rates are frequently encountered in areas of new public housing that would not be identified as inadequate in census surveys. Equally, it seems that multiple deprivation is more apparent in some inner-urban cores than in others (Hausner and Robson, 1985) – inner Birmingham, Liverpool, Glasgow and Manchester seemingly substantially worse than most others.

The scale of deprivation and the constraints on mobility within these cores should not, however, be exaggerated. Most heads of households within the cities do, on the whole, find employment, and job opportunities for women may be better than elsewhere. Wealth continues to be generated in the conurbations. A great deal of inadequate and over-crowded housing has been removed, not always to be replaced by high-rise developments. There is still some scope for the more deprived to move. A national mobility scheme operates within the public sector, and some social housing is constructed beyond the cities by housing associations. However, the central problem remains – cities may have lost segments of their populations, but they have been left with ever-more dependent communities.

Demographic Trends and the Cities

In some respects, socio-demographic trends appear to indicate further problems for the cities. For example, the proportion of elderly people living in the cities seems destined to rise, and many of these will be single-person households. This will increase pressures on social services and public-sector housing. Other groups, such as single-parent families, the poor and the disabled, will gravitate to or remain in the cities, partially as a result of the relatively greater availability of rented accommodation. The issue of rented accommodation seems likely to become much more problematic in the 1990s. The indications suggest that the Conservative government elected in 1987 is committed to reducing swiftly the scale of public-sector housing. It will be interesting to see the effect this has on homelessness. It will almost certainly place additional costs on local government generally, and urban authorities in particular.

Although the overall picture of socio-demographic change in the cities remains gloomy, it is, at the same time, important to stress the dynamic aspects of its nature. The most significant change in this respect is the slowing down in the 1980s of the decline in urban populations compared with the previous decade (Champion, 1987). For example, between 1971 and 1981 the eight principal cities of the UK were, on average, losing population at a rate of over 1 per cent per annum. From 1981 to 1985 the figure had fallen to about 0.33 per cent per annum, and London actually gained population in 1984, thus reversing a long period of decline. There may be a number of reasons why urban population loss has moderated. Higher birth-rates among ethnic-minority groups, urban house-building projects, the running down of the new-towns programme and perhaps an improvement in environmental conditions in the cities may all have contrib-uted to this. It is not, however, obvious that this will continue, and it is not clear whether it will benefit urban government. It may if more better-off households remain in the urban cores, but to what extent does this trend mean that even more disadvantaged groups will be trapped in the inner cities?

Urban Economic Change

There can be little doubt that one of the main reasons for the urban problem is a reduction in both economic activity and employment opportunities in the older conurbations. From 1977 to 1987, the scale of plant closure and/or contraction in many cities was extremely dramatic. In the late 1970s and early 1980s, most were losing well in excess of 1,000 manufacturing jobs per month. By 1987, the travel-to-work areas encompassing the seven major conurbations held more than 850,000 registered unemployed and, with the exception of Greater London, average unemployment rates were over 15 per cent (DE, 1987). In smaller pockets within the cities, and often in peripheral public-sector housing schemes, unem-ployment figures were much higher, with particularly disturbing youth unemploy-ment totals.

Increasing unemployment in the cities has had all kinds of unfortunate consequences: it reduces demand for such services as retailing, while at the same time increasing pressures on public-sector provision; and social services, housing, training and education encounter increased demands for certain activities. Rising unemployment also has implications for health provision and policing. It can lead to serious motivational problems, especially for the long-term unemployed. In short, unemployment must be considered as the primary agent causing and maintaining urban deprivation.

The causes of economic and employment decline in the cities have been the subjects of the most extensive studies – no other area of urban research has received so much attention. Here, where our main focus is on policy, some debates can receive only limited analysis. Nevertheless, we must consider some central issues, if only because associated policy has to be assessed in the light of the scale and intensity of the decline. Five major themes can be explored. They do not all explain the decline in total, nor are they mutually exclusive, but collectively they raise many of the prime considerations. These themes are the impact of national and regional economic performance on the cities; labour-market consid-erations; urban economic structure; spatial explanations; and the role of capital.

National and Regional Economic Performance

Cities inevitably reflect broader national and regional trends. In the case of national trends, fiscal and monetary policies adopted by the Labour government between 1976 and 1979 and particularly by the Conservative government between 1979 and 1982 had a marked effect on many urban manufacturing companies. Throughout this period, an increasing emphasis was placed on tight monetary policy in an attempt to eradicate inflation. Interest rates rose, Sterling increased sharply in value against most international currencies, public-sector capital investment was curtailed and VAT was substantially enhanced. The overall effect was extremely damaging for many companies in the cities and elsewhere: domestic demand was reduced; the costs of borrowing rose; public-sector contracts diminished; exports became much more expensive. Inevitably, industrial production declined. An index of 100 for industrial production in 1975 rose to 112 by 1979, but fell back to 100 by 1980 (British Business, 1981) and stood at only 109 by 1986 (British Business, 1988). Industrial investment was even more badly affected: in 1980 prices, UK industrial investment was about £8,200 million in 1979 but by 1987 was only £7,586 million (British Business, 1987). National economic policy does not explain the whole of urban decline, but the attitudes and policies of the Treasury have a greater influence on the cities than probably all urban projects combined. The over-simple assumptions widespread in the early 1980s that controlling the supply of money would in some way affect inflation (the removal of which would introduce a period of substantial growth) must be considered as being unsound.

Regional performance also influences the urban economies. Too often regional issues are neglected in an attempt to understand and to moderate urban decline. Recent research (Wolman, 1987) has confirmed that there is a strong relationship between regional and urban economic performance, not just in Britain but other countries as well. In the UK, the bulk of those cities and towns that performed well between 1971 and 1981, when assessed by such indicators as employment growth and unemployment, were located in the growth regions of the south of England. Almost all the poor performers were to be found in the economically-disadvantaged regions. Cities are not independent entities – they reflect national and regional considerations.

Urban Labour Markets

One reason for high unemployment in the inner cities is the concentration within them of groups of people who are prone to unemployment (Thrift, 1979). Among the most obvious categories here are the unskilled, the young, black people and those made redundant from manufacturing. These groups tend to locate in the older urban cores as a result of factors examined earlier. Many are trapped in the inner cores because of the unavailability of rented housing beyond the cities. These segments of society have been regarded as constituting one side of a dual labour-market (Lever, 1982). This argument proposes that some workers operate within an environment dominated by low wages, high unemployment and turnover rates, with limited opportunities for mobility. Their labour histories consist of periods of work interspersed with times of unemployment, under-employment or black-economy jobs. Their position can be contrasted with that of people in professional occupations, where unemployment is rare, mobility is relatively easy, jobs are advertised nationally and retraining is usual. An increas-ing proportion of the latter occupations seek to live beyond the cities and to commute back to them. Net inward commuting into the inner cities expressed as a percentage of total employment almost doubled between 1951 and 1981, from 21 per cent to 39 per cent (Begg, Moore and Rhodes, 1986). Frequently, urban employment rates fail to distinguish between those living and working in the cities and those commuting in. By 1981, two in every five urban jobs were occupied by commuters. Hence, an increasing proportion of the declining urban jobs (urban jobs almost halved between 1951 and 1981) went to commuters. By 1981, unemployment for those living in the cities was 50 per cent higher than the national average.

The quality of the labour supply in the cities might also be another important factor for high unemployment. There is evidence (Wolman, 1987) that urban economies with larger proportions of those skilled in complex manufacturing techniques and knowledge-based services tend to outperform those urban econo-mies with large proportions of people with lower skills in manufacturing. There may be a number of reasons for this. Skilled and professional employees may, perhaps, spend more on labour-creating services. They earn more. Evidence suggests a direct correlation between higher educational and technical qualifica-tions and the creation of new companies (Storey, 1981). The quality of the labour market may, therefore, affect economic development and unemployment in contrasting ways: an over-supply of lower-skilled labour will heighten unemploy-ment, while a higher concentration of the more skilled and better educated may well encourage greater employment, even if only marginally.

Urban Employment Structures

It could be argued that one reason why the cities have lost so many jobs and contain so many unemployed is that they are over-represented by sectors of the economy that have declined nationally – the cities merely reflecting national trends. Somewhat surprisingly, there is little evidence to suggest that a city's industrial structure in itself accounts for poor urban performance (Fothergill and Gudgin, 1982). Between 1952 and 1976, more than two million urban jobs were lost that cannot be accounted for in terms of national trends influencing unfavour-able urban structures (Danson, Lever and Malcolm, 1980).

An important issue when examining urban employment structures is that of the varying performances of different sectors of employment. It is necessary, for example, to distinguish between manufacturing employment and service employment, both in the private and public sectors. Wolman (1987) establishes that a strong relationship existed in the 1970s between poorly-performing urban areas and areas with a large percentage of their workforce engaged in manufacturing. This may seem inevitable as the percentage of people working in the UK in service employment rose from 49 per cent in 1963 to almost 64 per cent in 1983. The more successful cities might be judged as those able to attract and retain expanding service-sector employment, and that did not have an initially high proportion of manufacturing jobs.

In practice, however, the situation is not so simple. In the cities, public-sector service jobs declined by 7 per cent in the 1970s, and private-sector services fell by 22 per cent in the 20 years after 1961 (Begg, Moore and Rhodes, 1986). Equivalent national statistics show that public-sector jobs increased by 7 per cent between 1971 and 1981, and that private-sector services increased by 18 per cent between 1961 and 1981. Wolman (1987) concludes that there is little difference in service-sector growth between good and poor urban performers but that there are very marked differences in the degree to which cities lost manufacturing jobs. Some lost far more manufacturing jobs in the 1970s than others. According to Wolman, of the 15 per cent difference in employment performance between good and poor urban economies, three-quarters can be explained by contrasting experiences in manufacturing.

Not all analyses agree entirely with this conclusion. Hall (1987), for example – possibly because he uses different definitions and sample size – presents a somewhat contrasting picture. He assesses the degree to which manufacturing losses between 1971 and 1981 in eight major cities were balanced by changes in service-sector employment. All eight cities lost manufacturing jobs in the decade. By 1981, London, Liverpool and Birmingham had lost about 45 per cent of their 1971 totals. However, not all lost service-sector jobs. London, Liverpool and Glasgow lost some, but other cities including Edinburgh, Bristol and Cardiff increased their service-sector jobs by approximately 10 per cent and their information service jobs by 20–30 per cent. It was this balance or imbalance between losing manufacturing jobs, on the one hand, and service-sector performance on the other, that accounted for urban economic performance and variation in the 1970s.

Cities do not simply reflect their employment structures: they tend to perform worse than their economic constitutions would suggest. The amount of manufacturing employment in a city is important in the sense that those cities with a great many in industry may encounter the greatest loss of jobs. However, there are considerable variations between the cities. Some fare much better at retaining manufacturing jobs and creating service-sector jobs than others. There is also some evidence that cities doing well in manufacturing will also do well in service-sector employment. This certainly seems to apply to those cities that appeared to perform the best in the 1970s: Cardiff, Bristol and Edinburgh. Why did they do so well? As Hall (1987) concludes, possibly because of the decentralization of information-processing functions from London to Bristol and Cardiff, and because of the separate financial, institutional and political complexes in the provincial capitals of Cardiff and Edinburgh.

Spatial Explanations for Urban Economic Decline

A number of explanations for urban economic decline are based on the assumption that jobs and/or industrial investment have moved out of the cities because of prevailing production costs and the problems associated with the older urban cores. These explanations are treated at length in the work of Fothergill and Gudgin (1982) and Fothergill, Kitson and Monk (1985). This work attempts to explain the rural shift in manufacturing employment. Between 1960 and 1981, London lost 51 per cent of its manufacturing jobs, and the major conurbations 43 per cent, at a time when small towns lost only 1 per cent and rural areas actually gained 24 per cent. This marked urban–rural shift is explained by a number of causes. Although industry now requires fewer workers per unit of factory floorspace, its actual demands for floorspace have increased. While all areas of the country have lost employment as a result of the need for fewer workers, rural areas have gained in that they can accommodate new floorspace. Many older urban areas do not have large, easily-developed sites for new investment, and existing firms are often accommodated in inadequate, cramped premises. Once firms seek out new, more efficient premises, they tend to move to locations where production is more efficient and potential expansion easier.

In some cases this process may have been heightened by the imposition of regional policy discussed earlier. In London and Birmingham particularly, regional policy since 1945 may have hastened this decentralization of industry as industrial-development certificates were required from central government before larger manufacturing developments could be implemented. Of the applications, 90 per cent were approved but some – perhaps notably those relating to the car industry in the West Midlands – were not. This inevitably meant the dispersal of production away from the more prosperous areas of the south and Midlands towards largely green-field sites in the Assisted Areas of the north and west of Britain. The cities lost in this process. London and Birmingham lost manufacturing, and little of this went to other cities.

Although most of the debate surrounding the urban–rural shift has concentrated on the question of inadequate sites and premises in the cities, one allied development needing mention is the incubator thesis. Research (Fagg, 1980) has indicated that at least in some cities, inner-urban areas provide ideal factors of production for newer companies: cheap premises and labour can be obtained, markets are close and niches in production or distribution can be carved out. As some companies become more successful, they begin to develop internally those functions once external to their operations – development, sales, catering and other functions may be assimilated into the company, which inevitably requires larger premises. These functions, as mentioned earlier, are found increasingly beyond the urban cores.

The incubator thesis and other aspects of spatial explanations raise a number of issues. In the case of the incubator thesis, there is evidence (certainly from London – Nicholson, Brinkley and Evans, 1981) that few sections of industry seek out inner-urban locations, and that only a small number of companies migrate outwards in the classical fashion. This point is also significant in a broader context: spatial explanations are based on the assumption that the urban–rural shift is due primarily to company relocations in response to physical pressures. However, in most cities only a relatively small proportion of jobs lost go elsewhere, most are simply lost. Other explanations of urban employment change, therefore, need to be explored.

Urban Economic Change and the Role of Capital

Some of the most important contributions to urban economic analysis have come from those commentators who are determined to locate the declining output and employment in the cities within broader politico-economic considerations. Although some of these analyses have proved abstruse and others untenable, this collective contribution (which can be only briefly explored here) has proved vital in re-orientating thinking on the urban problem and its resolution.

Those working within the paradigm of capital and its effect on the cities relate economic decline to the workings of the market (those interested in pursuing the topic should consult the National Community Development Project (1977); Friend and Metcalf (1981); Massey and Meegan (1982); Scott (1982)). The debate can be summarized roughly as follows. Uneven development has always characterized economic activity in Britain and elsewhere. For a variety of historical reasons, initial development occurs in some places and not in others. Over a period of decades, however, the location of economic activity changes. Output becomes less efficient and profitable in the earliest developed regions as diseconomies of scale arise. Existing output and new production relocate, sometimes within different companies. This can be seen to underlie economic change in the UK in that the older industrialized regions have tended to suffer in relation to the expanding regions of the south of England. However, the picture is infinitely more complex than this.

The complexities lie with the processes of corporate restructuring. Britain has a particularly top-heavy industrial structure. By the late 1970s, about half of its industrial output came from its fifty largest companies, many of which were multinationals. These companies tended to be characterized by a series of activities:

1. They often acquired smaller competitors.
2. They sought out the most profitable lines produced in the most efficient of their plants operating with the cheapest labour available.
3. They tended to create complex company structures.
4. Small, prestigious head quarters remained in London.

5. Research and development centres were located in the shires.
6. Routine administrative functions were sited in the free-standing towns and cities within a hundred or so miles of London.
7. Production was possibly undertaken in the older industrialized regions, especially if projects could claim regional policy or some similar incentive.

Uneven development, and in particular the decline of the older industrialized regions, was therefore intensified by policies of corporate restructuring, which moved many non-productive functions to the south.

This brief summary only touches on what is a much deeper issue. Massey and Meegan (1982), in particular, have expanded the analysis in a number of ways. They point out that in practice there is great variety in corporate activities, even within one sector. Companies may intensify production, improve productivity or reduce output – all of which tend to reduce employment. In addition, the activities of management and labour can influence key decisions over investment and productivity, and these in their turn will have an influence on plant profitability. Nevertheless, despite these important reservations, it is clear that the kinds of policies effected by larger companies have been unfavourable to the older conurbations. Indigenous firms have been acquired by larger competitors, and production has often been abandoned or reduced. The local control of companies is lost. Industrial output is concentrated in the most efficient plants using the cheapest labour. Increasingly this is outside the UK altogether: international industrial investment undertaken by British companies amounted to over £10 billion per annum by the late 1980s.

The role of the State is not neutral here. Constraints on corporate activities become less limiting: planning controls are relaxed to encourage development in the south; regional policy is down-graded; international controls on capital are abandoned; and most takeovers are allowed by monopolies and mergers legislation, even when national and local benefits are not evident. When the State itself undertakes investment, it rarely does so to the benefit of the older industrialized regions. Notable examples include the siting of defence and research establishments to the south and west of London, in areas already attractive to the open market (Hall et al., 1987). It is this combination of corporate activities in an unfettered market, with limited State support for the older regions, that ultimately explains the scale of employment inequalities. While the older cities and regions lose jobs and output, the areas in easy distance of London gain employment, receive the benefits of State investment, and attract better-off people who create jobs for others in both the public and private sectors. These processes collectively impose unemployment rates of, for example, 4.5 per cent in Berkshire early in 1988, compared with 19.3 per cent on Merseyside (DE, 1988).

These perspectives on the urban problem – many of which have emerged from neo-Marxist debate – need to be treated with some caution. It could be argued, for example, that even if the analysis is correct, what would it matter? Markets have always sought out the most profitable locations, and this has inevitably created problems for declining areas. It is impossible to control market tendencies, and it is better to allow them to operate. Ultimately, the majority benefits – as the 1980s have shown. No matter how such thinking appeals to certain political and business interests, it is important to highlight the consequences of such attitudes for the older cities and the country as a whole. Industrial investment abroad is unlikely to create wealth and jobs to the same extent as would investment within the UK. Society cannot abandon extensive parts of the country – that would be socially unacceptable and economically untenable. Most corporate acquisitions achieve little for society as whole and tend to worsen economic ills in the older cities. The free market might provide the only politically-realistic model in which economic change can occur, but it still requires the tightest of legal and administrative controls.

The Urban Financial Context

Earlier sections of this chapter explored the inter-related effects of political, socio-demographic and economic trends on the cities. In this final section, brief mention will be made of that particularly complex policy area, urban finance. For many years, urban finance has been a system with which very few practitioners or academics have been totally at ease and procedures governing local-government funding have become more convoluted in the 1980s. The full operations and defects of this approach have been explored elsewhere, both in relation to current expenditure (Audit Commission, 1984; Travers, 1985, 1987) and capital expenditure (Jackson, 1984; Davies, 1987). Here we will highlight the impact, rather than the complexities, of the system.

By 1979, local-government current spending and manpower was at its highest level. The incoming Conservative government was determined to reduce this. Throughout the 1980s, the government tried to achieve this aim through a variety of measures. In 1979–80, the Secretary of State for the Environment, Michael Heseltine, reduced the rate-support grant (money provided by central government to fund local authorities) by £300 million. More fundamentally, in the early 1980s a new system of local-government support was introduced – the block grant. This attempted to establish uniform expenditure targets that, if substantially exceeded, would initiate penalties. By 1983–4, penalties on overspending authorities amounted to over £200 million – almost all of which fell on urban Labour councils. However, despite these controls and a general reduction in the proportion of local-government expenditure emanating from central government (down from over 60 per cent of spending in the 1970s to well under 50 per cent in the following decade) the Conservative governments elected in 1983 and 1987 decided to continue the process.

The 1983 government considered that the penalties incurred under the block-grant system had not done enough to curb local spending. It therefore implemented the Rates Act 1984, which allowed Parliament to restrict local rate revenues. It did this through ratecapping, a system that allowed central government to dictate rate-calls set by some or all local councils, and by joint boards governing areas such as fire, police and transport, established as a result of the abolition of the metropolitan counties. Thus, as Travers (1987) points out, by 1987–8 almost one-fifth of current spending by English authorities was under central-government control. This degree of control was introduced because the government believed that some urban Labour councils were profligate in their policies on spending – policies that would undermine the viability of local businesses. Rates, however, represent a relatively small proportion of total costs for most companies, and rate rises tend to occur because of external inflationary effects and not because of local-government policies (Midwinter, 1985). There also appears to be no relationship between high rates and unemployment (Crawford, Fothergill and Monk, 1985).

By the late 1980s, debates surrounding the rate question became largely irrelevant. For many years the Conservative Party had been committed to abandoning domestic rates, a commitment the 1987 Conservative government decided to uphold. In 1986 a green paper, *Paying for Local Government* (Cmnd. 9714) proposed to replace domestic rates by a poll or community tax, and to create a centrally-determined uniform non-domestic rate. The question of how local services should be funded had been the subject of a number of previous official reports, notably the Layfield enquiry (1976). In general, most of these commentaries had concluded that domestic rates were probably as fair and efficient a system as any. Despite this, however, a poll tax will be implemented throughout the UK by 1990.

Under a poll-tax system, all adults over the age of 18 will pay a flat rate set by the local authority concerned, although there will be rebates for the less well-off. One of the main reasons for the government introducing the system is to ensure that more residents pay for local services. The implication of this is that the more profligate councils will not be re-elected. However, only about one-quarter of total expenditure will be raised through the poll tax – the rest will come from a centrally-determined needs grant and a uniform business rate set by the government after a revaluation of non-domestic property. What are the likely implications of the new system for the cities? It is difficult to be precise until the system has been implemented, but some results seem likely. The cost of setting up and policing the system in the inner cities, with their greater proportions of rented accommodation, will be high. Many will not register and simply evade or try to evade the tax. Larger families, with which the cities are over-represented, will suffer relative to smaller households. The combined effects of the new central grant and the business rate seem likely to benefit suburban areas to the detriment of the urban cores (Huhne, 1988). However, the ultimate effect of the business rate seems unclear. Retailing is likely to suffer in relation to commercial and industrial property. Rate-calls for commercial and industrial property may fall for companies in the more depressed regions of the country (Centre for Local Economic Strategies, 1987b). This effect might be welcome in the cities. On the other hand, it is likely to be more than counterbalanced by heightened household poverty within the larger poorer families, and an overall reduction in local-government spending as a large proportion of expenditure will be controlled by central government.

Much of the debate concerning local government finance has concentrated on current, not capital, spending. But capital spending retains its own intrinsic importance. In the 1980s there has been increased central control (Davies, 1987). This strategy has attempted to moderate expenditure while disposing of public assets. It has not been a wholly successful policy. In the mid-1980s, many local councils substantially overspent their cash limits. This was due to a number of practices, including the willingness of local authorities to pre-fund or defer expenditure, and to enter into various lease-back arrangements with developers and/or funding institutions. However, this should not hide the fact that many, especially urban, councils have capital resources that are inadequate to deal with the problems that face them. The repair bill for public-sector accommodation is estimated to be at least £20 billion in 1986 prices (Association of Metropolitan Authorities, 1986), a year in which total capital spending amounted to about £5.5 billion.

With many of the creative accounting techniques implemented by councils in the 1980s being increasingly subject to central control, in the longer term public-sector capital investment looks very much like becoming a minority interest. This is somewhat ironic when it is realized that local government has substantial resources that cannot be used because of centrally-imposed prescriptions. Un-spent receipts from the sale of council houses, for example, were estimated at over £6 billion in 1987 (Page, 1987). This type of administrative regulation, when combined with a decline of one-quarter in current expenditure to the cities between 1981–2 and 1985–6, and the imposition of an administratively difficult and possibly inequitable poll tax, makes the financial environment within which the older urban cores have to operate singularly unfortunate.

Conclusions

This chapter attempts to raise some of the contextual issues that impinge on the inner-city debate. Clearly, not all of these can be illuminated. In many respects cities do no more than reflect the wider processes that prevail within society as a whole. Few of the constraints identified in this chapter apply solely to the cities. Processes of economic change, political marginalization, demographic restructuring and financial deficiencies affect communities that could not be identified as 'urban'. However, perhaps what does make the cities different is their scale of decline and deprivation: those seeking the most disadvantaged in contemporary Britain would do well to start in the older urban cores.

It is worth identifying those cities that appear to reflect most clearly those processes of economic decline, social deprivation and physical dereliction that epitomize the 'inner city'. Liverpool seems unique, not simply because of the scale of its problems but also because of the consistent pattern of political conflict that has characterized so much of what has happened in the city and in its relations with central government. Newcastle and Glasgow would appear to be in a somewhat better position, for although both have lost substantial numbers of jobs in manufacturing, a more optimistic picture has emerged in the late 1980s, based on public–private-sector investment and cultural developments. A third tier of cities might include Sheffield, Leeds, Manchester, Birmingham and parts of inner London. Aspects of inner-urban problems are evident in a range of smaller cities and large towns, such as Hull, Sunderland (Tyne & Wear), Barnsley (S. Yorks.), Nottingham, Leicester and many authorities within the North West and West Midlands.

There is an enormous variety in the urban experience that this book cannot hope to explore to any great extent: different cities have been subject to contrasting pressures and have responded in different ways. And, although the major conurbations have received much assistance through inner-city policy, relative rates of deprivation may be evident in much smaller communities. However, as the next part of the book shows, these smaller towns have gained relatively little from the urban dimension.

PART II

Inner-Urban Policy, 1977–88

3
The Labour Government and Inner-City Policy, 1977–9

As mentioned in Chapter 1, the publication of the 1977 white paper, *Policy for the Inner Cities*, signalled the emergence of permanent inner-urban intervention. However, a number of policy changes affecting the older cores had been introduced before that date. For example, changes were implemented in an attempt to ease the problems of London's Docklands, generally regarded as the largest area of urban dereliction in England. Local authorities were encouraged to advertise the advantages of the Docklands. This had hitherto been prevented in a desire to help encourage mobile industries to locate in the less prosperous regions of the north and west. However, by 1976, local councils in east London were able to publicize the advantages of their areas to industrialists.

Additionally, the government announced changes to the system of industrial development certificates. For many years industrial developments had required central-government approval, a mechanism designed to help employment relocate to the less prosperous regions, away from the South East and the West Midlands. Companies unwilling to move to the older regions had the option of moving to the new towns. From the mid-1970s, the Docklands were given equal priority with the new towns; after 1977 they were given a higher status.

Other policy changes designed to assist the cities were introduced before 1977. In the late 1970s, almost two-thirds of local-government current expenditure came from central resources, as determined by the rate-support grant. This varied between authorities according to the complex needs and resources formulae then in operation. There is evidence that for much of the Labour government's period in office between 1974 and 1979, the older urban cores were to receive greater support than other areas of the country (Jackman and Sellars, 1977).

In 1977, Peter Shore, Secretary of State for the Environment, announced that the new-town programme was to be curtailed (Shore, 1977). Since 1946, approximately thirty new towns had been built throughout the UK. They were intended to achieve a number of objectives. Some in the North East, such as Washington, were intended to act as regional growth centres. However, virtually all were designed at least partially to house people and jobs from the older urban cores. From 1977 onwards, though, the government decided to curtail the programme – a policy development the Conservative governments elected after 1979 were to endorse. No new towns were to be designated. Stonehouse, intended to relieve Glasgow's housing stress, was abandoned in 1976. Staff and resources allocated to it were redirected to a major project intended to regenerate east Glasgow.

The emergence of inner-urban policy in the late 1970s seems inevitably to have coincided with the abandonment of the new-town programme: the emphasis was clearly changing. The problems of the older cities were to be resolved – if resolved they ever were – on the spot and not through planned decentralization.

The 1977 White Paper: Policy for the Inner Cities

Policy for the Inner Cities (HMSO, 1977), remains the only white paper published on the problems affecting the older cities and the kinds of policies that might be introduced to overcome them. It identified a series of constraints impinging on the urban cores and on many of those living within them. Economic decline was the most serious problem, but it interacted with – and was partly responsible for – other aspects of urban decline: physical decay, social disadvantage and the particular difficulties affecting ethnic minorities. A range of policy innovations were needed to overcome or to moderate the urban problems. These would be undertaken by a number of organizations. Local authorities were seen as 'the natural agencies to tackle inner area problems' (*ibid*. p. 8), but local communities, voluntary bodies, the private sector and, where appropriate, central government were all to be involved in implementing policy.

The white paper outlined a number of policy innovations. Attempts were to be made to ensure that the cities would continue to do well out of the rate-support grant. The Location of Offices Bureau, which for many years had been attempting to move commercial development out of London, was given the task of promoting office employment within the cities, including London itself. Urban policies emanating from a number of central-government departments were to be co-ordinated. In retrospect, only three policy developments outlined in the white paper were important in the long run: the revamping of the Urban Programme, the Partnerships and industrial improvement areas.

The New Urban Programme

The Urban Programme was originally created in 1968, running in its original form for nine years. It funded many social and community projects in local authorities at a rate of 75 per cent. In 1977 its administration and function changed. It was substantially enlarged from £30 million per annum to about £125 million per annum, and was transferred from the Home Office (where it had been since 1968) to the Department of the Environment. One reason for this change was that the programme was to widen its coverage. Throughout the late 1960s and for much of the 1970s, the Urban Programme had concentrated on social schemes. After 1977 its brief was extended to include industrial, environmental and recreational projects – a reflection of the white paper's suggestion that economic decline was crucial to understanding the urban problem.

In 1977 the programme was changed in one other important aspect. Before 1977 there had been no formal ranking of the local authorities, and more than seventy had gained something from Urban Programme expenditure. However, inner-city funding introduced by Labour in the late 1970s began the process, which has continued since that time, of identifying and ranking authorities. This was made clear in the circular accompanying the Inner Urban Areas Act 1978 (DoE, 1978). Primarily, this Act provided local authorities with additional economic development powers. But, when taken in conjunction with procedures governing Urban Programme funding, it became apparent that three divisions of urban local councils, or 'designated districts', were to be established. At the base of the pyramid, 19 'other districts' in England and Wales were identified. In these authorities certain economic development powers were to be allowed. For example, 90-per-cent loans could be granted for the acquisition of land or the carrying out of works on land and industrial improvement areas (see below). At the intermediate level, 15 programme authorities were specified. These councils were granted the same economic development powers as the 'other districts', but they were additionally invited to submit an annual programme that would normally lead to greater Urban Programme funding than would be made available to the lowest tier of authorities. The most important developments, however, were to occur in the top tier of designated districts – the Partnerships.

The Partnerships

The concept of Partnership was first developed formally in the 1977 white paper. The government argued that 'if real progress is to be made in tackling some of the major concentrations of problems, special efforts must be focused on a few cities in the next few years' (HMSO, 1977, p. 16). Therefore, within certain urban cores the government was prepared to offer special arrangements: it was willing to enter into formal arrangements with the then two tiers of local government to devise and to implement urban strategies. Central government's commitment was important, as it would help to instil confidence in the areas concerned, and would assist in unifying the actions of both central and local government.

Although the government was placed under considerable pressure by many of its own back-benchers to widen the concept to other areas, only seven Partnerships were created in England: the Docklands of London, Hackney–Islington, Lambeth, Newcastle–Gateshead, Manchester–Salford, Liverpool and Birmingham. With the possible exception of Lambeth, these were not surprising selections. In Partnership areas, a committee was created consisting of representatives of local and central government, the health authorities, the police, voluntary groups, and so on, chaired by a central-government minister. In most Partnerships, some form of Inner City Unit was created, to service the key committee and to co-ordinate the annual Partnership Programme.

The Partnership Programmes began to appear in 1978, tending to conform to a standard pattern. Economic problems were often identified as important, although other constraints were seen to affect many of the deprived households. Little analysis was apparent in many of the early Programmes – a defect that was never entirely overcome. Instead, the Partnership Programmes consisted largely of lists of social, economic, recreational and other projects for formal Partnership approval, which was usually forthcoming subject to the total availability of resources. Despite the somewhat bureaucratic and unimaginative nature of the early Partnerships, their creation stimulated interest, and a series of research reports on their initial development emerged. The comments and criticisms made can be categorized into two substantial debates – funding and management.

Partnership Funding

A number of investigations were undertaken into the scale and direction of Partnership funding during the late 1970s (Hall, 1978; Nabarro, 1980; Lawless, 1981a). It was evident that total resources allocated to the initiative were never great. Of the first portion of enlarged Urban Programme funding of £125 million for 1979–80, the Partnerships received approximately £66 million. This was not evenly split. Docklands, for instance, received £15 million, and Lambeth £5 million. In addition, a number of one-off payments were made to the Partnerships. They received £15 million from Operation Clean-Up (an environmental improvement project) and about £70 million from a capital projects' scheme announced in the 1977 Budget. Derelict-land grant-allocations were also biased towards the Partnerships.

Even if all of these funds are taken into account, total direct funding to the Partnerships in the late 1970s has to be regarded as small. Total funding for economic development projects in Liverpool for one year through the Urban Programme was less than Unemployment Benefit paid in the city in one week. To some extent the spending was diversionary because it deflected attention from the cuts that were occurring in the orthodox funding mechanisms for current and capital expenditure. Between 1974 and 1979, Liverpool's capital programme was reduced from almost £60 million to less than £40 million.

It was always assumed that Urban Programme funding would represent only a proportion of total expenditure in the Partnerships – most would come through a diversion of mainstream budgets to the benefit of the older urban cores. However, there is little to suggest that this occurred. Although it is not easy to establish exactly what proportion of total local spending goes to specific areas of the cities, where efforts were made, the inner-urban areas do not appear to have done particularly well. Research done in Liverpool – albeit before the Partnerships – (Nabarro and McDonald, 1978), indicates that the inner areas received less than might be expected considering their total population. Similarly evidence for housing in the North East suggests that housing expenditure was not biased towards the Partnership areas (Butler and Williams, 1981). It is extremely difficult for local government to discriminate positively in favour of certain areas. More money for some areas means less money for others, which can obviously raise political issues. As a large proportion of current expenditure goes on salaries (particularly in education) there may be little scope for re-allocating resources. The Partnerships did not find themselves suddenly in receipt of substantial resources deflected from other local-authority budgets.

What of central government? Is there any evidence that the Partnerships obtained additional resources from the bending of appropriate central-government expenditure sources? It is more difficult to trace the spatial patterning of central-government spending than it is at a local level. It is, however, possible to make reasonable estimates of where some budgets were allocated (Lawless, 1981b).

In the late 1970s, two budgets that were potentially important in urban economic development were the regional programmes implemented by the Department of Industry, and the policies effected by the National Enterprise Board. These programmes were identified by the government (HMSO, 1977) as central to urban regenerative strategies, but neither programme was redirected successfully towards the cities in the last few years of the 1974–9 Labour administration.

In the case of regional development grant, for example, the three Partnerships in the assisted regions actually received less than their late 1970s industrial populations would have suggested. On the whole, they were not favoured in the allocation of regional selective assistance where there was an element of discretion on the part of the Department of Industry. This pattern was repeated by the National Enterprise Board (NEB). The NEB was created by the government in 1975 to help restructure British industry by selective intervention, particularly in areas of high unemployment. However, by the end of 1978 only 12 companies in the whole of the North West and North East of England had received financial support, amounting to a total of little more than £3 million. This apparent neglect of the cities in the late 1970s by central-government agencies needs some qualification. Certain programmes, for example, some MSC schemes and the construction of industrial property by English Estates (a Department of Industry subsidiary) were beginning to favour some cities, especially Liverpool and Newcastle. But there was no abrupt re-orientation of government spending towards the urban cores from 1977 to 1979.

There may have been reasons for this lack of positive discrimination towards the older conurbations. In some cases, such as the regional development grant, there was little scope for re-allocation as spending was demand-led. It was difficult to redirect rapidly the thinking of the Department of Industry, which had traditionally been interested in regional, not urban, policy. Inertia no doubt played a part. Within the NEB it was clear that a regional – let alone a city –

dimension was never to play a very important role in determining policy. Not a great deal of intervention took place, and most that did was directed towards small companies in 'high-tech' sectors, which tended to be located in the south of England. Although it might have been hoped that local and central government would redirect main programmes to the benefit of the cities, there is little to suggest that this occurred. At least initially, additional resources to the cities were largely contained within the Urban Programme.

Although inner-urban funding in the late 1970s was limited, the direction of Urban Programme expenditure remains a valid area of investigation. A great deal of the spending went on capital projects. This was understandable, because it was far easier to direct capital projects to defined inner areas in very visible – and therefore politically appealing – projects. The difficulty was that capital schemes often required current funding if they were to operate. For example, social and community centres required staff. As Urban Programme allocations were limited, covering only 75 per cent of total costs, capital projects did not always receive appropriate current support.

Some areas of spending received relatively little through the Urban Programme. Education and, to some extent, housing, were allocated only limited support, partly because they were seen as dominating mainstream local-authority expenditure. All of the seven Partnerships allocated at least 15 per cent of spending to economic development projects in the late 1970s, and similar proportions were spent on environmental schemes. The social, community and recreation budgets received more. Both Lambeth and Hackney–Islington, for example, granted these areas at least 40 per cent of Urban Programme spending in 1979–80.

The early allocations of Urban Programme spending might appear somewhat surprising when the emphasis of the inner-city drive in the late 1970s was towards jobs and economic development. This was not reflected in Partnership spending patterns. This apparent discrepancy can be explained in a number of ways. There were difficulties in rapidly-increasing capital economic projects. Much of this spending went on new industrial units, schemes that might take years to implement, and for which there may have been an appropriate upper limit. Established departments, such as social services, possibly dominated early Partnership discussion. The more powerful, well-established organizations were often able to put forward their own projects – many of which might simply have been cancelled because of capital cut-backs after the International Monetary Fund imposed constraints in 1976. The initial allocations of Partnership spending in the late 1970s were often the result of the large, better-prepared and more politically-aware departments putting forward their cases more persuasively than the others.

This rather pragmatic approach towards Partnership spending allocations received unfavourable comment. Hambleton (1981), for example, argues that the late 1970s saw a shift in policy towards a programme-dominated approach. In policy strategy, which had been well developed in the Inner Area Studies, Community Development Projects and the 1977 white paper, broader issues of substance had dominated the debate. The complexity and inter-relatedness of urban problems had been stressed, existing policies had been evaluated and a series of policy modifications had been proposed relating to the economy, housing, population dispersal, urban administration, and so on. But once the Partnerships were formally created, the emphasis had switched to a programme-dominated approach. Issues relating to the efficacy or the alignments of new or existing policies were not discussed. Much of the debate was about the allocation of relatively small sums of money to specific projects.

The partnerships' essentially pragmatic approach inevitably meant that important debates were never really aired. For example, there may have been little point in devising an 'inner area' and allocating resources to it. Many of the urban deprived did not live in the 'inner area', and urban economic processes operated across entire city regions. The narrow focus on an area also deflected attention away from such issues as economic and demographic mobility within and between the cities, and the problems associated with many public-sector housing schemes built on the edge of cities in the 1960s and 1970s. Some indicators, such as unemployment, may have classified these developments as being at least as deprived as the inner-urban areas, but they did not gain from the inner-city initiative. Partnerships tended not to address issues such as training, education, the needs of the elderly, the problems of single-parent families, race and policing. The assumption seemed to be that through some mechanism or other, the additional allocation of relatively small sums to be used for capital projects would moderate the problems of the inner-urban areas. It is hard to understand how this might have occurred, especially when the primary cause of urban economic decline – the processes of corporate restructuring discussed in Chapter 2 – were never addressed.

The Management of Partnerships

Financial considerations proved to be only one of the problems facing the Partnerships, the other major constraint relating to the question of management. The approach was based on notions of co-ordination and co-operation. Representatives from as many relevant agencies as possible met a few times a year, and devised an agreed strategy for the inner area concerned. The Partnership itself had no executive power. Policy modifications and developments occurred through consensus and persuasion. Not surprisingly, this approach met with formidable operational difficulties.

Not all central-government departments were convinced of the need for an inner-area strategy. Between 1974 and 1979, for example, the Department of Industry allocated about £250 million in regional assistance to Merseyside – almost all of which went to green-field locations (Nabarro, 1980). Why should that department suddenly have directed its policies to the benefit of the inner city? Inevitably there were problems of intra- and inter-institutional conflict and jealousies at a time when three tiers of administration – the city, the county and

central government – were all involved. Different organizations operated from
different political standpoints, from within different boundaries and on different
time-scales. To one observer of the Lambeth Partnership, 'the implications for
future corporate behaviour within and between statutory agencies are hardly
glimpsed' (Spooner, 1980, p. 105).

The creation of the Partnerships had considerable implications for the voluntary sector. The old Urban Programme, run from the Home Office until 1977,
tended to benefit the voluntary sector. A great deal of spending went on social and
community projects, an area where the voluntary sector flourished. The reshaping of the Urban Programme, the creation of Partnerships and the specific
emphasis on economic projects meant there was a real danger that the voluntary
sector would be left out. The Secretary of State for the Environment argued,
however, that the voluntary sector would have a genuine role to play (Shore,
1978). Certainly this happened in some Partnerships. In Lambeth, for example,
about one-quarter of expenditure in 1979–80 went to the voluntary sector, which
was represented on appropriate committees (Spooner, 1980). In Newcastle there
was a distinct attempt to involve voluntary and community groups within a
decentralized management structure (Beecham, 1978).

Elsewhere, tensions occurred between the voluntary sector and local government (Green, 1978; Liverpool CVS, 1978). In some cases decisions were taken
with little regard for voluntary interests. In some Partnerships, voluntary-sector
funding amounted to less than 10 per cent of total Urban Programme expenditure. In most cases this was a substantial decrease compared with community
funding undertaken by the old Urban Programme from 1968 to 1977. The early
development of the Partnership concept appeared to hold out little encouragement for the non-statutory sector. However, with the election of a Conservative
government in 1979, questions relating to the organization of Partnerships
became of secondary importance: the important issue was whether Partnerships
would be retained.

Partnerships, 1977–9: Conclusion

Partnerships represented the first major inner-city initiative in England. They
emerged as a result of substantial research undertaken by a series of urban
experiments operating for the best part of a decade. At the time they appeared to
offer the greatest opportunities for the cities, yet ten years afterwards they appear
to reflect a lost age. Partnerships were based on the assumption that reasonable
people working in a multitude of organizations could devise an agreed strategy
for parts of urban Britain: local and central government would work in harness
and would incorporate other interests – the police, the business community,
voluntary groups, and so on, into a coherent administrative whole. Consensus
and co-operation would ensure success. The reality of some of these assumptions
is debatable. The political climate has changed so rapidly since 1979 that many of
the central principles of the Partnerships have been severely eroded. Centrally-

determined control, commitment and conviction replaced consensus and co-operation.

The Glasgow Eastern Area Renewal (GEAR) Project

Partnerships were not created in Scotland; instead, in 1976, the Secretary of State for Scotland announced that a planned new town at Stonehouse would be abandoned, and that staff and resources were to be directed towards a major regenerative project based in east Glasgow. The Glasgow Eastern Area Renewal project – GEAR – was to become one of the largest, if not *the* largest, urban-renewal project ever attempted in Europe. A range of organizations was involved in the implementation of the scheme, notably regional and district councils, the Scottish Special Housing Association and the Scottish Development Agency.

These organizations were to initiate a renewal scheme for one of the most deprived areas in Britain. East Glasgow's population had fallen from about 100,000 in 1961 to 45,000 in 1976. There had been an enormous loss of jobs, and many of the younger and more mobile had left – sometimes voluntarily and sometimes through planned dispersal – to peripheral public-housing schemes.

The early development of GEAR was not without its problems. The administration of a project involving so many organizations proved complex (Leclerc and Draffan, 1984). The Scottish Development Agency (SDA), which had been given the task of co-ordinating the various bodies involved, was not over enthusiastic about the project (Morison, 1987). However, unlike the position in England – where the Partnerships controlled so few resources directly – the SDA had its own funds to help expedite the strategy.

This strategy was based on a number of key objectives. In some cases these have clearly been achieved: housing policies, for example, appear to have been successful (MacLennan, 1987; MacLennan, Munro and Lamont, 1987; Morison, 1987). Almost 6,000 older properties were renovated between 1977 and 1982 and, by 1985, over 3,000 new private-sector dwellings had been started or planned. These policies have helped to stem population loss and have begun the process of social change designed to encourage more middle-class households to live in east Glasgow. In addition, environmental standards have improved – sometimes dramatically – and leisure and recreational facilities have been substantially enhanced.

But a number of criticisms have been levelled at GEAR. For example, it has been criticized for being unable to deal with problems of unemployment, which continued to rise throughout the 1980s. Manufacturing employment, for instance, fell by 40 per cent between 1979 and 1982. Training facilities have been marginal to local needs. Although a great deal of economic development has occurred, with at least 150 new or refurbished units constructed by 1982, the total net number of new jobs for local residents has been very small, and about three-quarters of the companies operating within the area have relocated from elsewhere (McArthur, 1987). Moreover, there is little to suggest that public-sector investment has been successful in encouraging equivalent private spending. Of the £400 million spent in GEAR by 1985, 60 per cent was by the public sector (Morison, 1987).

The most severe criticism of GEAR has come from Booth, Pitt and Money (1982). They argue that a great deal of public-sector investment would have inevitably occurred, and that little of it was new. In many respects the management of GEAR has been ineffective: goals have been defined too broadly, simply to ensure consensus; organizations have been assimilated within GEAR for purely cosmetic purposes; power has been too diffuse; and too much has been made of environmental improvements when the local economy has been subject to dramatic changes and unemployment has risen substantially. Whatever the merits of these criticisms, it seems reasonable to conclude that this model of urban regeneration based on the co-ordination of public-sector organizations by a lead agency is unlikely to be used soon again. It may have had a certain validity, especially in Scotland with its constellation of key organizations within or close to Glasgow. However, changing political circumstances – notably the downgrading of public-sector intervention and the emergence of centrally-appointed development agencies – make any repetition of the GEAR model improbable.

Industrial Improvement Areas

The third major initiative introduced by the government in the late 1970s were the Industrial Improvement Areas (IIAs). The 1977 white paper, *Policy for the Inner Cities* (HMSO, 1977), argued that additional powers should be granted to local authorities to enable them to assist industry. A number of suggestions were outlined in the paper, many of which were formalized in the Inner Urban Areas Act 1978. These included powers to make loans for land acquisition and works within the designated districts; loans for site-preparation works; and grants for rent relief within Partnerships. However, the most important power was that all designated districts could declare Industrial Improvement Areas.

The concept of improving older industrial areas was not new. Authorities such as Rochdale (Grt Manchester) and Tyne & Wear had used existing legislation to declare informal IIAs before 1977. There were a number of reasons why authorities began to look more carefully at older industrial property. Much of it had disappeared during the redevelopment programmes after 1945. Comprehensive redevelopment usually involved the removal of mixed land-use areas (often with a number of buildings dedicated to industrial production) and their replacement with property used for aspects of consumption, for example, housing. This process often involved the compulsory acquisition of firms and hence their potential closure. Many companies, however, used their compensations to re-capitalize (Chalkley, 1979). Nevertheless, older industrial property could be useful for new companies wanting cheap premises and wishing to locate in the cities.

Considerations such as these provided the rationale for the formal creation of

IIAs in the 1978 Act (DoE, 1978). They were declared in run-down industrial and/ or commercial areas, and were seen as a means of stabilizing economic activity. Once declared and approved, authorities were advised to concentrate a range of activities on the IIA, including planning, traffic management and land-assembly functions. The areas needed to be selected carefully. Some industrial areas were too delapidated while others needed no support. Central government assumed that typical IIAs would be up to 50 hectares in size, and that the programme would last about 3–5 years. To assist in the implementation of IIAs, local authorities were given powers to provide loans and grants for improving amenities, and grants for converting or improving buildings.

By 1984, over two hundred IIAs had been declared, covering about 7,000 acres. Almost all of the designated districts had declared at least one improvement area, and about one-quarter of declarations were in commercial or commercial/ industrial areas. Most funding had come from the Urban Programme, amounting to about £9 million per annum. Other sources of funding included the Derelict Land Grant and MSC resources.

There have been a number of critiques of the IIA concept (Tym, 1977; Topham, 1978; Etherington, 1987). These culminated in a major review undertaken by consultants for the Department of the Environment (JURUE, 1986a).

In some respects, IIAs proved relatively successful. There has tended to be an increase in confidence within IIAs, although this has rarely been matched by increasing private-sector investment. Some, but not many, jobs have been created by companies receiving assistance. Under Section 6 of the 1978 Act, which provides support for improvements to buildings, this amounted in 1984 to about £11,000 per job – which was relatively cost-effective. Costs per job in schemes where developers had received assistance to improve property for use by others were much lower. In practice, however, these represented a small total number of posts. Some environmental improvements and land reclamation had additionally taken place.

It is apparent, however, that the IIA approach has produced a number of problems and complexities. There is little to indicate that environmental and/or property improvements actually create jobs, boost output or encourage invest-ment. Support does not necessarily result in employment creation. Difficulties related to access, parking and outmoded infrastructures often remain. Improve-ments to rented property may raise rents to the detriment of the producers as opposed to the landowners. Too much emphasis may have been placed on grants at the expense of loans for environmental improvements (Meyer, 1986). Loans allow local authorities to make a profit and will impose commercial considera-tions on companies having to repay assistance. The costs of building refurbish-ment may be greater than is often assumed. The approach, however, is fundamentally marginal: it is not going to have a great effect on urban economies faced with the impact of massive corporate restructuring (see Chapter 2).

However, even if it is accepted that IIAs are marginal to the prospects of urban economic development and employment creation, it seems evident that some approaches to industrial improvement are more successful than others. For example, some authorities have selected inappropriate areas, have not fully appreciated local business trends and opportunities, have undertaken little publicity and have not encouraged participation by companies in the implementation of the programme (Meyer, 1986). Equally, not all authorities have managed to co-ordinate the totality of their activities relevant to the operation of industrial improvement. However, it is possible to link improvement to wider conservation polices, to develop a range of financial packages appropriate to individual firms and to integrate IIA declarations with land-use planning (Etherington, 1987). Even where councils are more effective in initiating and developing IIAs, the real effect of such activities can easily be overstressed.

The Labour Government and Inner-City Policy, 1977–9: Conclusion

It may seem harsh to criticize the Labour government for its inner-urban policy at the end of the 1970s – there was little time for the introduction of a comprehensive programme of action towards the cities before the political realities of the 1979 election. However, Labour's urban policy cannot be perceived as anything other than meagre. Marginally, more resources went to the Urban Programme, but the cities lost funds from mainstream expenditure. The Partnerships proved a largely bureaucratic device, with little influence and only minimal funding. IIAs were given legal status but their impact was slight. In effect, as McKay and Cox indicate (1979, p. 255), 'no sensible observer inside government or out could expect the limited measure announced in 1977 to transform the inner city'. The whole exercise was administratively top-heavy, with little in the way of innovation. It could be seen as an attempt by the government to try to retain urban seats or, perhaps more accurately, as the reflection of a personal commitment towards the inner cities on the part of Peter Shore, Secretary of State for the Environment (MacKay and Cox, 1979). There is evidence for this in, for example, *The Castle Diaries* (Castle, 1980). The election of the Conservative government in 1979 resulted in the appointment of another Secretary of State for the Environment, who had altogether more radical ideas for inner-city intervention.

4
Conservative Inner-Urban Policy after 1979: The Search for Co-ordination

The Conservative governments elected in 1979, 1983 and 1987 initiated a number of inner-city policies. By the late 1980s, *Action for Cities* (HMSO, 1988a), identified at least twelve separate interventions, although some of these were insignificant in resources terms and not all observers would regard some of the programmes as intrinsic to the inner cities. Nevertheless, the period after 1979 undoubtedly saw a proliferation of inner-city projects. In this and the next two chapters, we attempt to unravel and evaluate policies effected by the Conservative governments after 1979.

There are a number of possible classifications of urban intervention after 1979, including (most obviously) a chronological approach; here, however, we adopt a different approach. Three main strands to Conservative policy can be discerned from the multitude of programmes that were either introduced after 1979 or that, although initiated before that date, were perpetuated in a modified form by subsequent Conservative administrations. One major theme, already mentioned in the previous chapter, is the search for co-ordination, a search that has figured prominently in a number of urban programmes. This is the main focus of attention in this chapter. Two other important issues have also framed inner-urban policy in the 1980s. One of these – deregulation and liberalization – is evaluated in the next chapter. Chapter 6 examines the development thrust within Conservative urban policy.

This thematic classification is not without its drawbacks. Inevitably individual urban initiatives do not always fit solely within one or other grouping: there is a degree of overlap between the classifications. Nevertheless, this threefold approach has its advantages. It provides a means by which urban initiatives can be grouped into coherent themes and it highlights the underlying ideas on which so many of the programmes are based.

Co-ordination after 1979

Just as the search for co-ordination was a central rallying call for urban projects and experiments in the 1970s – with the creation of the Comprehensive Community Programmes and Partnerships – so the 1980s saw similar developments. In 1988, for example, *Action for Cities* (possibly the most important statement on the cities made by a Conservative government after 1979 – HMSO, 1988a), highlighted the government's determination to pursue the theme of integration. It argued that public-sector resources were substantial, 'but efforts need to be pulled together more effectively, and brought to bear in the same place at the same time' (*ibid*. p. 4). In the press conference accompanying the 1988 launch, the Prime Minister made a statement reiterating this argument: *Action for Cities* was not about new resources nor new policies, but about co-ordination and about building on apparent strengths (Platt and Lewis, 1988).

This concept of improved co-ordination has emerged in a number of urban projects. These can be categorized into three main areas of debate: the Urban Programme and the Partnerships; the City Action Teams and Task Forces; and other co-ordinating initiatives.

The Urban Programme and the Partnerships

The Conservative government elected in 1979 inherited a number of urban policies that had been initiated by the Labour government between 1977 and 1979, and that are discussed in Chapter 3. As that chapter points out, the underlying principles of the Labour government's approach in the late 1970s – partnership between central and local government, and consensus and co-operation – became much less fashionable in the following decade. Nevertheless, Conservative governments elected in 1979 and after continued to support the major urban drive developed in the late 1970s: the Urban Programme and the Partnerships.

This continuity in policy can be explained in a number of ways. The Urban Programme and the Partnerships were relatively non-controversial, funding many worthwhile projects of a social, environmental and economic nature. In addition, the Partnerships allowed central government a say in previously locally-determined issues. In any case it is always easier to initiate, rather than terminate, policy. In 1979, after eleven years of existence, the Urban Programme was particularly characterized by a network of local-government officials and councillors, voluntary-sector representatives, civil servants and a range of other interested organizations forming a mechanism through which valuable, if somewhat stereotyped, projects might be funded. However, radical governments, such as that elected in 1979, might not be swayed by such considerations. Indeed, there were a number of reviews of inner-city policy (for example, DoE, 1984). None of these resulted in the demise of the Urban Programme, not least because after 1979 central government introduced a number of policy modifications.

The modifications introduced should not be regarded as fundamental. The 1988 Urban Programme cannot be seen as dramatically different from the approach inherited by the Conservative government elected in 1979. But there have been changes. In 1981, the government re-affirmed its emphasis on economically-orientated projects within the Urban Programme – a trend made much more probable in that, particularly within the Partnerships, local Chambers of Commerce were invited to comment on programme submissions. Throughout the 1980s there was also a gradual change in the list of qualifying councils. By 1987, 55 Urban Programme authorities had been designated (see Figure 1, p. xi) (DoE and DE, 1987).

By 1987, 57 local authorities were requested to submit Inner Area Programmes which were to outline the objectives of Urban Programme funding, and to indicate how other funding programmes dovetailed into Urban Programme activities (DoE, 1989). Of these 57 authorities, 32 had been Partnership or Programme councils, and the remainder had a lower status either as Other Designated Districts or benefited from the traditional Urban Programme from which very many councils had gained some, often limited resources. The stratified approach to Urban Programme funding, with in essence a fourfold division of councils – Partnerships, Programme Authorities, Other Designated Districts and traditional Urban Programme councils – was effectively to be abandoned. Instead a single Urban Programme designation was to be introduced, and the traditional Urban Programme was to be phased out. This approach was adopted in order to concentrate resources on areas of greatest need (Young, 1986a). In the event in 1987–8 all 57 authorities submitted an Inner Area Programme and all were approved.

Two other changes in the implementation of the Urban Programme after 1979 can be identified. First, a more managerialist tendency is apparent. One problem that had consistently undermined the Urban Programme was its lack of objective definition (National Audit Office, 1986). Partly in response to this criticism, the government introduced in 1985 the Urban Programme Management Initiative (DoE, 1985c). This provided a series of standardized measures against which to evaluate individual projects, while at the same time laying down broader objectives for the whole programme. These objectives were to improve employment prospects in the inner cities; to reduce derelict land and sites; to strengthen the social fabric of the inner city; and to reduce the number of people in acute housing stress.

Second, it has been argued (Stewart, 1987) that one major (if implicit) trend within the Urban Programme was its determination to utilize inner-city intervention as a mechanism through which to address problems associated with black communities. To Stewart (*ibid.* p. 134) for example, it appears that

whilst there remain widely varying explanations of the reasons for the disturbances of 1981, 1983 and 1985, there can be no doubt that the fear of social disorder, and its association with the needs and demands of black people, has been a major factor in sustaining central government interest in inner cities policy.

The Urban Programme has been one policy instrument through which the government has been able, if only at the margins, to influence the position of black people in the major conurbations.

Bearing in mind the policy modifications imposed on the Urban Programme after 1979, what kind of supportive environment was it providing for the cities by the mid-to-late 1980s? In financial terms, it wasn't a particularly large programme: by the mid-1980s it amounted to about £300 million. In 1986–7, for example, it stood at £317 million. Of that sum, the Partnerships received £123 million. However, the tradition of allocating some Partnerships substantially more than others had been perpetuated: Birmingham received £24.5 million while Lambeth received £10 million. In all the Partnerships, more than half the expenditure went on capital as opposed to current spending and, in some areas such as Manchester–Salford, the capital-expenditure figure was much higher.

There was in 1986–7 considerable variation between Partnerships in the balance of expenditure. In the 1980s, the proportion of spending on economically-orientated schemes, for example, constructing small industrial units, rose in the Partnerships and in the Programme Authorities from 29 per cent of Urban Programme expenditure in 1979–80 to 38 per cent by 1984–5. Liverpool, however, even by 1986–7 was allocating less than 20 per cent to economic projects, although the equivalent figure for Newcastle–Gateshead stood at 43 per cent. Environmental projects (for example, clearing derelict sites) and housing expenditure remained relatively insignificant in all the Partnerships except Liverpool, where 40 per cent was spent on these two areas. On the other hand, the largest single allocation in virtually all the Partnerships was for social schemes, which were usually concerned with leisure projects or with providing services for the disadvantaged. At least 40 per cent of expenditure was usually directed towards social projects, despite the supposed economic orientation of the Urban Programme.

This apparent discrepancy is to a certain extent explained by the relatively large sums in the Partnerships going to the voluntary sector, which is normally stronger in social rather than in environmental or economic areas. In 1986–7, in most Partnerships at least one-quarter of expenditure went to the voluntary sector, and in Birmingham, Lambeth and Hackney the figure was much higher. About £15 million of a total Partnership expenditure of £123 million was spent on projects likely to benefit ethnic minorities.

Patterns of expenditure in 1986–7 in the second tier – the Programme Authorities – were not significantly different from those in the Partnerships. Slightly more was spent on economic projects and slightly less on social schemes. However, rather more was usually allocated to capital projects: at least 60 per cent expenditure was so defined, with almost 90 per cent going to capital schemes in some councils. The main point to make about the Programme Authorities was that their total allocations from the Urban Programme were very small. Cities as large as Sheffield and Leeds were receiving no more than £4 million per annum, little more than 1 per cent of their annual current expenditure. Nevertheless, to receive their Urban Programme funding, they had to produce an annual programme for submission to the Department of the Environment.

In the late 1980s, the annual programmes produced by both the Partnership and Programme Authorities were in many respects not substantially different from those produced a decade earlier. In line with the programmes produced in the late 1970s, a brief contextual statement would be followed by a closer definition of the objectives to be pursued in the economic, social, environmental and housing areas, and the role to be played by various statutory authorities and the voluntary sector (Birmingham City Council, 1987b; Newcastle City Council, 1987). The bulk of the programme would then outline in some detail the specific projects being funded, and the resources allocated to them over a 3–4-year period. The number of projects could be surprisingly large: for example, over 850 were listed in Birmingham's 1987–90 programme, of which 218 were new.

Although the programmes devised in the late 1980s were not dissimilar to earlier statements, there were some variations in emphasis. The later programmes laid down clearer guidelines about the objectives and role of Urban Programme funding than was always apparent in the later 1970s. Many objectives were rather bland – perhaps an inevitable reflection of (in the Partnerships at least) an executive committee comprising such varied interests as central and local governmental, district health authorities, the police, the probation service and chambers of commerce and industry. However innocuous overall objectives might appear, there can be little doubt that a much tighter evaluation of Urban Programme funding had been implemented since the 1985 review (DoE, 1985c).

In the statements presented in the later 1980s, a whole series of forecast outputs frame expenditure categories. Economic objectives, such as improving or converting buildings for industrial use, are assessed against measures that include units created, total floorspace involved and numbers of jobs created and/or preserved. While no doubt imposing a degree of justifiable administrative rigour on Urban Programme expenditure, the attainment of such measures can too readily be equated with 'success'; obtaining accurate assessments of some outcomes, notably job-creation totals, is fraught with difficulties (as examined in Chapter 9.)

More revealing insights into the operations of the Urban Programme can be gleaned from an examination of the Partnership's role rather than through explorations of anodyne objectives. The Birmingham Partnership, for example (Birmingham City Council, 1987b), argues that Urban Programme funding cannot solve the problems of inner Birmingham. The Partnership can, however, play a distinctive role in promoting the effective direction of both public- and private-sector resources towards the most disadvantaged areas. In particular, the Partnership can help to lever funds from other organizations into inner-city investment projects; it can assist in the creation of innovative approaches towards inner-urban renewal; and it can operate as a last-resort funding for worthwhile projects that would not otherwise be implemented. As the following discussion points out, not all observers believe these roles have been consistently applied to the Urban Programme.

The Urban Programme and the Partnerships after 1979: a Critique

For the first fifteen years or so of its existence, little was published in the way of critical assessment of the Urban Programme. However, from the mid-1980s onwards, a series of research projects emerged on the consequences of the programme, particularly within the Partnerships. With more than 12,000 separate projects funded by the Urban Programme in 1985–6, some positive assessments of the approach can inevitably be anticipated. Some reviews of aspects of urban intervention indicated that Urban Programme funding often proved crucial for the implementation of projects as diverse as business-development initiatives (DoE, 1988b) environmental improvements (JURUE, 1986b) and some employment-creation projects (DoE, 1986).

One particularly illuminating evaluation of the impact of the Urban Programme in the mid-1980s was the four-year review of Birmingham's Partnership (Aston University, 1985). Some positive findings emerged from this assessment. The Partnership had helped to stabilize conditions in the inner city, partly as a result of encouraging a range of organizations to focus attention on the older urban core. Most projects had been efficiently run, and many had helped disadvantaged groups. Both households and companies within the inner-urban area felt greater confidence in the inner city as a result of Partnership policies. These had often operated most effectively in smaller areas within which housing or industrial-renewal projects could best be implemented. Interestingly, a later internal annual review of Birmingham's Partnership concurred with some of these findings and also identified other positive outcomes (Birmingham City Council, 1987a). For example, the Partnership had encouraged the creation of multi-agency working parties to explore such issues as crime, it had redirected funds from other agencies and the private sector through policies including industrial improvement and conservation, and it had helped to foster such new projects as promoting credit unions.

But, despite positive evaluations of the Urban Programme and particularly its operation within the Partnerships, it is evident that the whole strategy is open to criticism. These criticisms can be divided into three major areas: administration, funding and policy development.

Urban Programme administration

The fundamental administrative problem has been the perennial difficulty in achieving a co-ordinated response to the problems facing the inner-urban cores. These constraints were, of course, apparent when Partnerships were created in

1977: attempts to bring together different tiers of government and different departments of either local or central administration have proved consistently complex to implement. A particularly scathing review of urban initiatives in Merseyside (undertaken by a House of Commons environment committee) makes this clear (House of Commons Environment Committee, 1983). Different tiers of government refused to co-operate; Liverpool City Council disbanded inner-city working parties and incorporated Urban Programme issues into mainstream committees; and the Secretary of State steadily distanced himself from what he perceived to be a largely meaningless 'talking shop'.

Problems of co-ordination probably reached their worst by the mid-1980s. The disappearance of the metropolitan counties, and the adoption by many urban Labour councils of a rather more pragmatic approach towards central government, eased some of the particular difficulties facing the Partnerships. Nevertheless, the need to develop an action programme from a diverse range of interests has undoubtedly encouraged an overly bureaucratic philosophy to dominate both Programme and Partnership Authorities (Stewart, 1983; Parkinson and Wilks, 1986; Stewart, 1987). Too much emphasis was placed on the necessary function of actually producing an annual programme to which different parties can agree and that identifies the full list of projects seeking funding. While such an exercise can be useful in highlighting and co-ordinating inner-city initiatives, it is usually time-consuming relative to the resources on offer (Aldridge and Brotherton, 1987).

Urban Programme funding

The limited size of Urban Programme resources has been discussed above; for many Programme Authorities, Urban Programme funds are insignificant. However, for the Partnerships more is at stake. All the cities have lost resources as a result of reductions in central-government support from the block grant: between 1981–2 and 1985–6, block grant to the cities declined by one-quarter in real terms, and by 1985–6 penalties imposed on the Partnership and Programme Authorities in the rate-support grant settlement almost equalled their Urban Programme allocations (Association of Metropolitan Authorities, 1986). In the 1980s, cities have not gained in resources from central government.

Despite the limited size of the Urban Programme, the administration of grant has been consistently problematic (West Midlands County Council, 1986; Stewart, 1987). Limitations on virement ensure that the annual budgetary cycle becomes vitally important. Spending resources before financial deadlines has come to dominate the administration of Partnerships and Programme Authorities, instead of wider strategic issues. The increasing emphasis on capital projects within an annual budgetary cycle creates problems in that it may take a number of years to acquire and to develop urban sites. Some capital projects obviously need current resources to run them. Government approval has been increasingly required for minor projects, which normally cannot generate income. Finally, there is the contentious issue of time-expired projects. Many schemes are funded for three to five years and, although some tapering mechanisms may allow continued but limited funding thereafter, in the long run local authorities may have to bear a larger proportion of total expenditure if schemes are to continue. This has tended to happen, but declining local-authority current expenditure makes this practice increasingly unlikely – certain worthwhile projects will disappear because of a lack of resources.

The Urban Programme and policy development

Perhaps the most important criticisms levelled at the Urban Programme relate to policy development. A number of assessments argue, for example, that the Department of the Environment has traditionally not been able to provide strategic guidance governing the role and direction of the Programme (National Audit Office, 1986; House of Commons Committee of Public Accounts, 1986b). Until the 1985 major review of the Programme, objectives were ill-defined or absent, and broader issues of policy development and evaluation were neglected. There was little monitoring until the later 1980s (Sills, Taylor and Golding, 1985). The need for authorities to satisfy bureaucratic procedures generally stifled more important debates relating to, for example, the role of the inner cities, the position of the disadvantaged, the relationship between the inner cities and surrounding city regions, and so on.

Although it was always assumed that the Urban Programme would aim to initiate innovative projects, there must be some doubt whether this has always happened. In the late 1980s there is little to suggest originality in the lists of projects supported by Partnership and Programme Authorities. Economic projects tend to support companies and infrastructural provision; social and community projects usually involve funding for community and leisure centres or educational and advice projects; housing schemes are directed towards improvement and the needs of special groups; and environmental expenditure is allocated to improving older buildings and vacant sites. Little in this might be described as innovative. Most is common ground for local authorities and other agencies. Some initiatives may have been initially funded through Urban Programme expenditure: the enveloping of the external fabric of older dwellings and managed workshops (discussed in Chapter 7) were initiated using the Urban Programme, but it would be difficult to argue that these would not have occurred if Urban Programme funds had not been available.

One can see why the Urban Programme has not, on the whole, fostered a great deal of innovation. Departmental priorities have often featured prominently. For some departments in local councils, the Urban Programme has been regarded as a way in which capital schemes cut from other mainstream budgets might be reinstated. Urban Programme funding has, therefore, all too often operated not as a last resort, supporting imaginative projects, but as a substitute for resources cut from other budgets. The administrative structure of Partnerships, with its heavy emphasis on co-ordinating different agencies, would in any case tend to inhibit the unorthodox. Certainly there is little indication of policy innovation or dynamism in the Urban Programme. While presiding over the implementation of a large number of separate projects the entire structure of the Urban Programme seems increasingly marginal to the problems of both urban decline and main-stream inner-city policy. It might have moderated things at the margins, but it has done little to address or solve issues of economic decline, deprivation or inade-quate housing in the cities. Will it survive long into the 1990s?

City Action Teams and Task Forces

The City Action Teams' (CATS) origins lie in the government's response to the 1981 riots in Liverpool. Although many British cities saw riots in that year, the disturbance at Toxteth was the most significant in the context of inner-urban policy. This was because the Secretary of State for the Environment, Michael Heseltine, was charged by the Prime Minister to take particular interest in, and suggest responses to, the problems of Liverpool. To assist in this exercise, Heseltine, temporarily 'Minister for Merseyside', created a supportive Task Force. This consisted of about thirty full-time civil servants, mainly drawn from the Department of the Environment but with secondees from the Departments of Industry and Employment. At the same time, Heseltine also created a Financial Institutions Group, consisting of secondees from major companies who were asked to review the role of, and constraints upon, private-sector investment in the cities. Some of these companies visited Liverpool in 1981 at Heseltine's request, and as a result some private-sector managers were invited to join the Task Force.

It is clear that, when Heseltine originally returned to the Cabinet from Liverpool in the autumn 1981, he was intent on expanding public-sector invest-ment in Merseyside (Parkinson and Duffy, 1984). The Treasury defeated this proposal, but the Cabinet was prepared to initiate a new approach to Liverpool's difficulties – the Task Force. Cabinet's vision of the Task Force was at variance with Heseltine's own ideas. The role of the Task Force, as specified by the Prime Minister's Office, was essentially a strategic vision: broad issues were to be raised; the direction of any additional expenditure was to be stipulated; and longer-term policies were to be highlighted. In the event, Heseltine encouraged a much more pragmatic approach, centred on project development. Large-scale, politically-visible, catalytic schemes – such as the Garden Festival and Wavertree Technology Park – came to dominate the Task Force. Strategy never figured prominently except in the most generalized way. The private sector, for example, was central to the approach, and improving the city's image was seen as vital. The Task Force, however, was not really about broader strategy and policy it – was very much the creature of the Secretary of State and, like him, it was pro-active, project orientated and anti-bureaucratic (Morison, 1987).

There have been a number of evaluations of the Task Force (Parkinson and Duffy, 1984; Morison, 1987). The approach brought central-government departments into closer practical contact with problems, and thus encouraged civil servants to adopt more pro-active as opposed to supervisory stances. It helped to develop valuable projects; it utilized market skills in the implementation of schemes; and it began that process of integrating central-government departments into corporate entities through which regional problems might be better addressed.

But the Task Force highlighted some of the intrinsic problems inherent to programmes designed to co-ordinate agency response. Strategic analysis and vision was almost totally absent. It failed to create a genuinely corporate approach towards the problems of a single, major conurbation. The Department of Industry was consistently lukewarm to the concept of favouring one particular inner-urban area. The Task Force brought no specific resources of its own. It had to proceed through persuasion and encouragement – a difficult process with public-sector bodies, but almost impossible with major companies. Having successfully seconded through the Task Force one manager to refurbish industrial estates in Knowsley and Kirkby, United Biscuits promptly closed a food-processing plant at Knowsley with the loss of 2,000 jobs. There were problems in the relationship between the Task Force and local authorities. Eventually, formal administrative structures were created between the Task Force and local government, but the Task Force's determination to implement specific projects rather than to help create a broader and longer-term strategy for the conurbation inevitably created tensions.

It seems in retrospect that the Task Force in Merseyside was driven by Heseltine's particular vision of urban regeneration. The cities would not be re-invigorated by public-sector inspired and resourced strategies. Instead, he promoted a 'Disraelian sense of the obligation of local businessmen to exercise leadership in the big cities as their predecessors had in the Victorian heyday' (Parkinson and Duffy, 1984, p. 81). Once Heseltine departed, succeeding Secretaries of State for the Environment such as King, Jenkins and Baker, were unwilling or unable to sustain this vision. By 1983 the Task Force had become more institutionalized in that it had begun to deal with matters relating to mainstream local-authority funding programmes, Urban Programme submissions, derelict-land grant applications, and so on.

Even if the political status of the Task Force within Merseyside began to wane after Heseltine's departure from the Department of the Environment, one central objective of that initiative was to be retained: the co-ordination of central-government activity within the major conurbations. To achieve this, it was announced in 1985 that five CATS were to be created in the Partnership areas of England. One would deal with the three London Partnerships, and there would be others established in Birmingham, Liverpool, Manchester–Salford and New-castle–Gateshead. In 1988 two other CATS were announced based in Leeds and Nottingham. These initiatives were designed to co-ordinate spending within the Partnerships from the key departments – the Departments of the Environment, Employment, and Trade and Industry – spending that by 1988–9 amounted to about £850 million per annum. They were designed to operate closely with local government and the private sector to bring about the economic development of the inner areas. As with the Merseyside Task Force, they were to be primarily concerned with project development and not with issues of strategy.

Little substantive evaluation of CATS has emerged, perhaps not surprisingly when one bears in mind their location within central-government departments. Under their aegis a range of economically-oriented projects has been effected, including training schemes, science parks, support for small businesses, industrial-improvement projects, and so on. Anecdotal evidence suggests that CATS have not sought out local-government co-operation: they have implemented little that would not in any case have occurred. And, most importantly, they have been unable to devise anything that might be termed a corporate central-government strategy towards inner-city areas. One reason for this may be that the civil service's centralized administrative structure, based on line departments, makes the task of developing an integrated strategy in the English regions exceptionally difficult to achieve.

In part to complement the broad-based approach towards co-ordination developed within the CATS, the Inner City Initiative was launched in 1986 to increase private-sector investment in small urban areas that were subject to high unemployment rates. To achieve this, eight (later sixteen) Task Forces were created. They were located throughout England (see Figure 2, p. xii) and included Chapeltown in Leeds, St. Pauls (Bristol) and North Kensington. They consisted of small groups of civil servants and industrial secondees who attempted to initiate modest proposals to reduce local unemployment, in part by linking government investment in such areas as public works and training, to private-sector developments. For example, they tried to link training placement to jobs, or to target construction jobs on local people. Most have small-business development funds for new companies. However, their budgets are small: the 1988 *Action for Cities* programme granted them £20 million. It seems likely that they will make little impact, one reason being that they appear to be more concerned with targeting existing jobs rather than creating genuinely new ones.

Other Co-ordinating Projects

Mention should be made of some of the other urban projects that are designed to improve agency co-ordination. Not all of these approaches are necessarily directed solely to the cities, and some would not feature in many classifications of inner-urban policy. However, their potential implications for the major conurbations merits their inclusion here.

Three projects can be mentioned. In 1988, *Action for Cities* announced the development of a Safer Cities scheme, designed to reduce urban crime. Between 1988 and 1991, in 20 urban localities, representatives from the police and probation services, local businesses, local community groups, and so on, are to identify particular problems that increase local crime and to suggest appropriate remedies. In housing, there is the Priority Estates Project that, since 1979, has shown how the management of run-down estates can be improved by local authorities and tenants working together (Power, 1987). In training, *Action for Cities* identified a number of initiatives that were designed to improve access to and awareness of training services for those in the cities. Improved management techniques could include new information points in the inner cities, a decentralization of some services to deprived areas and the appointment of specialist staff to deal with the problems encountered by the long-term unemployed.

Co-ordination in Inner-City Policy: Conclusions

In this and the previous chapter, policies designed to improve the co-ordination of urban agencies have been addressed. The overwhelming conclusion emerging from any assessment of these programmes is that they have only a marginal impact on inner-urban problems. There may be circumstances where an integration of different programmes can reduce duplication and sharpen policy effectiveness, but in general co-ordination achieves very little – it is extremely difficult to put into practice. It assumes that an easy consensus can be achieved about the nature of the problem and its resolution, when this is rarely the case. It assumes administrative efficiency will suffice when this may only have the most limited of practical effects. It relegates crucial issues, such as equity and strategy, to minor roles. It gives priority to project implementation when policy development is needed. Finally, it almost invariably assumes that substantial improvements can be obtained with the same resources. In brief, it doesn't work. If it does achieve anything, it diverts attention from central issues of policy, expenditure and strategy. It is surprising to see it figuring still as an important mechanism through which urban policy might be delivered when, as the next two chapters show, other approaches have made a much greater impression on urban development. However, it may be wrong to assume that this inability to create a co-ordinated response is a bad thing: a truly-integrated response to the cities from the Conservative governments elected in 1979 and after might have been even more detrimental to the major conurbations than it has been.

5

Conservative Inner-Urban Policy after 1979: Liberalization and Enterprise

In the previous chapter we examined the strand of thinking and policy development that is centred on co-ordination. A second important theme in the evolving urban debate after 1979 is the issue of deregulation and decontrol. Conservative governments elected since 1979 have argued consistently that one reason for Britain's traditionally weak economic performance, when compared with other developed economies, has been the degree of regulation, control and taxation imposed on producers and potential entrepreneurs. This theme has run through white papers (HMSO, 1985; 1986a), statements from central-government ministers (Rumbold, 1986), and has increasingly impinged on the urban debate. The Prime Minister, for example, writing in *Action for Cities* in 1988 argued that 'the Government has created a climate which supports enterprise and has set about removing obstacles in the way of inner city recovery' (HMSO, 1988a, p. 2).

The truth of these assumptions − that Britain is a particularly regulated economy where enterprise might flourish in the wake of decontrol − is by no means held universally. These debates are developed in Chapter 9, which provides an overview of the entire range of inner-city interventions. Here, we outline and evaluate those urban initiatives that can be seen to have resulted from the drive for deregulation and decontrol. Three policies merit consideration: enterprise zones, Freeports and Simplified Planning Zones.

Enterprise Zones

Although there has been a proliferation in inner-city innovations in the 1980s, there can be little doubt that enterprise zones, together with Urban Development Corporations (see Chapter 6), have been the most significant of initiatives. Two major debates surround enterprise zones: there is the whole question of their origin, development and administration, which raises a series of issues in relation to governments' ability to bring about radical change; and, second, and more importantly by the late 1980s, their impact can now be assessed in some detail.

Enterprise Zones: Origins and Development

Enterprise zones originated in a number of ideas that began to amalgamate towards the end of the 1970s. They reflected a concept that had been current for a number of years, that experiments in planning free zones ought to be devised (Banham *et al.*, 1974). Those advocating such innovations were motivated most powerfully by aesthetic considerations, in particular by the deadening effect on design of development control. One of those promoting a liberalization of planning regulations, Professor Peter Hall, developed his ideas of deregulation into what was to become an influential contribution (1977).

Hall's ideas, which he was later to endorse strongly (Hall, 1982a; 1982b) were based on the idea that inner-urban decline in some British cities had gone so far that orthodox approaches to regeneration would not work. He argued instead for the establishment of a small number of what he termed 'freeports', within which unregulated free enterprise would be encouraged. The zones would be outside the effective control of national government. There would be free movement of goods, labour and capital. Wage levels would not be controlled and personal and corporation taxes would be kept to a minimum.

Hall justified his approach in a number of ways. Establishing such enclaves would inevitably encourage the inmigration of larger companies seeking tax exemptions. And so, overall, few new jobs would be created. Hall argued, however, that even this least satisfactory outcome would be worth while if it meant some of the long-term unemployed in the inner cities obtained jobs, and were thus able to climb up the skills and income levels. Those losing jobs from companies relocating to the inner cities would probably be able to acquire new jobs much more easily than those trapped within the cities. More optimistically, Hall argued that the zones would be about enterprise, growth and innovation. In some of the Pacific-rim economies, such as Hong Kong, unregulated enterprise had been associated with substantial economic progress since 1950 that, somehow, had resulted in real technological development. On an experimental basis at least, Hall argued the effectiveness of liberalization within a few inner-city sites should be explored. It could mean that some of the technologically-advanced jobs developing or locating in the Pacific might emerge instead in the UK.

These ideas, based on liberalization and decontrol, struck an immediate chord in the Conservative Party. Mrs Thatcher's election to Party Leader in 1975 was to be immensely significant in the context of post-war politics: the tradition of consensual, one-nation government (which had dominated Conservative governments since 1951) was, if not to disappear, to become far less dominant. Market-orientated, almost *laissez-faire* attitudes figured ever more prominently in the Conservative Party when in opposition in the 1970s. Once in government in the late 1970s and early 1980s, Hall's deregulatory message was taken up with alacrity.

This attitude was most apparent in a speech given in the Isle of Dogs in 1978 by Sir Geoffrey Howe, later to be Mrs Thatcher's first Chancellor (Howe, 1978). He argued that enterprise zones should be established in which detailed planning controls would cease to exist, certain legal obligations on employers such as employment protection legislation would no longer apply, various taxes would not be levied and the overall management of zones would be undertaken not by local authorities but by some other agency. These ideas formed the basis of enterprise zones that received formal approval in the Finance Act 1980 and the Local Government, Planning and Land Act 1980.

As a number of observers have pointed out (Taylor, 1981; Morison, 1987), enterprise zones differed in important respects from the outline statement made by Sir Geoffrey Howe in 1978. The idea was examined initially by the incoming Conservative government in the Industrial Policy Group at the Treasury. As a result of these considerations, the government decided that enterprise zones would be implemented by local authorities and not by some outside agency. Employment protection legislation was to be retained, and tax havens were eventually ruled out because of the problems of potential abuse by incoming companies or firms moving some plant into the zones. But certain exemptions would apply within a select and small number of zones. Three issues are important in this context: land and planning considerations, financial incentives and zone designations.

Land and planning

Howe had argued in 1978 that planning controls had, to a certain extent, been responsible for urban dereliction. Too much emphasis had been placed on supervising minor residential modifications and in rejecting applications, which had resulted in reduced economic growth. Enterprise zones, therefore, introduced the concept of the simplified planning regime, intended to speed up planning applications and to give deemed planning consent for many developments. Provided certain standards, such as height, access and land use were met, planning approval would not be required. These kinds of exemptions were regarded as potentially harmful by organizations such as the Royal Town Planning Institute. They would encourage the reincarnation of a Victorian urban Britain in which environmental standards would decline (RTPI, 1979).

These fears proved largely groundless. As the first set of declaration reports were produced, it became apparent that development within the zones would not be significantly different from what would have occurred anyway (Corby District Council, 1981; Newcastle upon Tyne, 1981; Swansea City Council, 1981): planning controls were often retained along zone boundaries; Special Industrial Uses, including noxious and dangerous processes, were still subject to control; and, on occasion, environmental improvements were written into declaration reports. Enterprise zones have not been environmentally harmful; in many cases standards have greatly improved.

The most contentious planning issue was perhaps the question of retailing. Some local authorities expressed a concern about the potential implications of removing planning controls from shopping, on the grounds that the financial incentives (discussed below) would attract into the zones large retail organizations, thus undermining existing shopping patterns. Certain authorities, originally identified by Sir Geoffrey Howe as potential recipients of enterprise zones in the 1980 Budget speech – notably Sheffield and Wolverhampton – were unable to reach agreement with the Treasury over this issue, and were thus not granted zones. However, the government accepted that large hypermarkets would still require planning permission, and in some cases (for example, Clydebank and Newcastle), it was agreed that declaration reports should include clauses governing shops over a certain size.

The question of land ownership and development also became somewhat complex. In his 1978 speech, Sir Geoffrey Howe said that new agencies would be created for the zones, which would re-allocate public-sector land to private users. As Morison (1987) makes clear, this did not occur, and the public sector was vital in the early development of many of the zones. In Dudley (W. Midlands), where a large proportion of land was privately owned, the Department of the Environment insisted that if enterprise-zone designation was to go ahead, it would only be approved if private-sector land was first sold to the local authority before a certain date. The most successful zones were those where one public-sector development organization, such as the local authority or English Estates, owned most of the land and was able to undertake necessary infrastructural investment. Too often, where land ownership was fragmented, development was slow to occur: statutory bodies were reluctant to sell unwanted land, and small, private owners did not often have the expertise and resources to benefit from financial incentives.

Financial incentives

Although the simplified planning regime had, on the whole, a mild effect, some financial incentives, on the other hand, undoubtedly stimulated development within the zones. The financial and business exemptions applying to enterprise zones in 1980 were rather eclectic. They included, for example, exemptions for industrial development certificates and relief from development land tax, both of which were to become irrelevant when these regulations were abolished everywhere. Firms within enterprise zones were also exempt from industrial training-board levies – hardly a massive concession, since the number of boards was to be substantially reduced throughout the 1980s.

Two concessions, however, proved much more important. Firms within enterprise zones would not pay rates for ten years, local government being reimbursed for lost revenue by the Treasury. This was an important incentive. Rates are frequently a small proportion of total outgoings, but there is little doubt that they are unpopular within the business community. The other concession was that there would be 100 per cent capital allowances on both industrial and commercial buildings. This was essential to the development of many zones, but not necessarily as had been anticipated. It was an incentive that applied particularly to developers and developer/producers, but not to those simply producing goods and services. It especially benefited commercial development. It had already been possible to offset some of the costs of industrial building, but this incentive was now extended to commercial projects in the zones.

Zone designations

It is important to stress that enterprise zones were regarded as a political experiment by the Conservative government elected in 1979. Ministers ostensibly believed that economic progress had been limited by control, regulation and over-taxation. Enterprise zones were a way of testing the viability of these ideas by liberalizing the market. As we have seen, some of the radicalism inherent in Howe's 1978 speech was dissipated by the time enterprise zones were given legal status in 1980. Nevertheless, observers on what might broadly be called the left were generally very much opposed to the creation of enterprise zones (Anderson, 1980; Massey, 1982).

The zones would apparently diminish employment protection and welfare legislation; they would create few, worthwhile, genuinely new jobs; and they diverted attention away from the dramatic contraction in investment and employment that was occurring in large parts of Britain in the early 1980s. Many Labour councils held these kinds of sentiments at the time. In the early 1980s, there had been a noticeably leftwards swing in some urban authorities, and to some councils the idea of upholding an initiative apparently rooted in the *laissez-faire* capitalism of the nineteenth century was unthinkable. On the other hand, protracted financial limitations were in force, and enterprise zones offered the hope of some financial support from central government.

It was the hope of central-government support that, on the whole, dominated thinking within Labour councils. Enterprise zones were simply not going to reflect uncontrolled capitalism. Although in early discussions surrounding enterprise zones a few authorities were unable to reach formal agreement with the government, most Labour authorities, with varying degrees of enthusiasm, eventually sought designation. By 1985, 25 enterprises zones had been established in the UK (see Figure 2, p. xii). Eleven were designated in 1981–2, and a further 14 in 1983–4. Extensions to 3 of the first 11 zones – Dudley, Wakefield (W. Yorks.) and Swansea – and to one of the second 14 zones – N.W. Kent – were made between 1983 and 1986. In 1988, a further (and possibly final) zone was declared at Inverclyde. Zones have been declared throughout the UK: 2 in Northern Ireland, 3 in Scotland, 3 in Wales and 17 in England. About half of these are in urban localities, with designations in Hartlepool (Cleveland), Dudley, the Isle of Dogs, Rotherham (S. Yorks.), Middlesborough, Tyneside, Speke (Merseyside), Salford–Trafford, Wakefield, Swansea, Tayside and Clydebank.

Political considerations governed the selection of some sites. Corby (North-ants), for example, received a zone after the closure of the British Steel plant, and an associated rise to over 25 per cent in male unemployment. In general, it was apparent that as far as possible different regions of the country (with the exception of East Anglia and the South West) were to gain at least one designation.

The zones varied significantly in size and composition. Some were little more than 50 hectares, although the Tyneside zone was over 450 hectares. Many were scattered over separate sites. This was perhaps predictable, since the Department of the Environment, in approving designations, wanted to declare vacant or derelict land as enterprise zones, thus avoiding the incorporation of existing companies, which would benefit from financial incentives, into designations. Some zones, such as that at Clydebank, consisted of one large site. Details of the physical characteristics of the zones, however, are of secondary importance to the central question – how has the private sector responded to these limited experiments in deregulation?

Enterprise Zones: Impact and Implications

One of the somewhat stranger requirements of companies in enterprise zones was that they would have to provide central government with only minimal statistical information. One wonders how many firms believe this requirement has actually operated, for no urban initiative has ever been the subject of so much formal and informal evaluation. For the first three years, consultants monitored the zones (for example, Tym and Partners, 1984). The Department of the Environment has itself undertaken a number of reports updating progress in the enterprise zones (DoE, 1985b; 1987a). The National Audit Office (1986) and the House of Commons Committee of Public Accounts (1986a) have produced their own evaluations. Some local authorities have published assessments (Swansea City Council, 1986; Trafford Metropolitan Borough, 1988). Independent observers have published widely on the zones and their effects (MacLeary and Lloyd, 1980; McDonald and Howick, 1982; Norcliffe and Hoare, 1982; Catalano, 1983; Talbot, 1988). A major evaluation was undertaken by P.A. Consultants for the Department of the Environment and published in 1987 (P.A. Consultants, 1987).

This mass of material can be grouped into a number of key areas. In the following discussion, four major debates are isolated: the jobs issue, economic development, planning and property implications, and costs and perceptions of incentives.

The jobs issue

Jobs have always been central to the debate surrounding enterprise zones. In ways never always explicitly developed, early proponents of the idea argued that incentives within the zones would help sustain employment (Hall, 1977; Howe,

1978). This might initially involve low-grade jobs relocating from elsewhere into the zones. However, Hall (1982a; 1982b) hoped that higher-skilled and better-paid jobs would eventually appear through the same processes that had driven countries such as Hong Kong, South Korea and Singapore from low-grade, low-paid economies into higher, if not always 'higher-tech" production.

The reality has turned out to be somewhat more prosaic. By 1986 there were approximately 63,000 jobs in the 23 British enterprise zones. This gross figure needs considerable refinement. Consultants evaluating the zones suggest that only 55 per cent of these jobs are net additional jobs in the zones (P.A. Consultants, 1987). Moreover, these net additional jobs are offset by reductions in jobs in surrounding areas – by a diversion of activity into the zones, and a displacement of jobs in firms out of the zones. These job losses in off-zone areas total almost 30,000. However, off-zone areas benefit from the creation of local construction jobs, and through multiplier effects: the expenditure available to new employees in the zones percolates through the local economy; and migrants attracted to the area create additional demands for jobs in housing, education, and so on. The culmination of these processes suggests that about 12,800 additional jobs have been sustained either directly or indirectly by the enterprise-zone experiment.

Evidence is available relating to the occupational structure and characteristics of the jobs created within the zones. Not surprisingly, the older and more industrialized zones have higher proportions of male employment. In total, almost 70 per cent of jobs in British enterprise zones are taken by men. Of posts available within the zones, 16 per cent are professional, 13 per cent clerical, 32 per cent skilled and 39 per cent unskilled. These totals inevitably vary considerably from zone to zone, depending on its economic structure, location, and so on. For example, 30 per cent of the jobs in Dudley are professional, but only 7 per cent of those in Wellingborough (Northants). Whereas Salford–Trafford has 17 per cent skilled posts, the total in Scunthorpe–Glanford (Humberside) is 60 per cent.

Economic development

Using the information from the major review published in 1987 (P.A. Consultants), various aspects of economic development within enterprise zones have become more apparent. By 1986 there were just over 2,800 establishments in the 23 British zones, 70 per cent of which were located in the first designations of Swansea, Tyneside, Corby, the Isle of Dogs, Wakefield, Dudley, Speke, Clydebank, Salford–Trafford and Hartlepool. About 50 per cent of the companies in the zones are engaged in manufacturing, 7 per cent in retailing and 16 per cent in distribution, although there is considerable inter-zone variation. Almost 90 per cent of firms in the Corby zone are in manufacturing, while over 25 per cent of Swansea's firms are in retailing. Only about 12 per cent of companies might be termed 'high-tech', although this figure should rise as a higher proportion of post-designation companies fall into this classification. As they increase in proportion to the pre-designation firms, so the number of 'high-tech' jobs should increase. Almost 90 per cent of companies have fewer than fifty employees.

An important debate surrounding the concept of enterprise zones has been the question of displacement. Would existing firms relocate into the zones because of their incentives, to no overall effect? By 1986 it had become evident that these fears had, to some extent, been realized. About 34 per cent of employment within the zones was in pre-designation companies, and about 34 per cent had transferred in. New start-ups provided 14 per cent, and new branches 18 per cent. Of the employment in new branches, about 50 per cent had relocated from elsewhere within the UK, 35 per cent from the local area or the region and about 14 per cent from abroad. In relation to transfer, about 50 per cent of employment had come from the local area and almost 25 per cent from the region. Transfers from elsewhere within the UK amounted to 18 per cent and only a tiny proportion had come from abroad. Most activity in the zones in 1986 was clearly either there pre-designation, or it represents movements in by existing branches or by firms within the region.

Another issue of considerable importance is the question of additionality: what proportion of companies have located on, or remained within, the zones as a result of designation, and what proportion would have located there without designation? This issue raises a whole series of complex methodological questions that are examined in some detail in P.A. Consultants' 1987 review. Total additionality for pre-designation firms amounts to 23 per cent of companies. The zones have thus had some success in preventing the relocation or closure of pre-existing companies. For post-designation companies, about two-thirds would not have been in the zones had designation not taken place. For new start-ups, the figure rises to over three-quarters. When analysed by sector, it is the manufacturing firms that show the highest additionality. It is also clear that additionality has varied considerably amongst zones: while Rotherham's additionality amounts to 60 per cent, others, such as Dudley and Clydebank, reveal totals of about 30 per cent.

One final issue in relation to economic development is the question of the performance of companies in the enterprise zones. On the whole, the evidence (P.A. Consultants, 1987) suggests that companies located in enterprise zones have performed better than firms located elsewhere in the local economy. This seems to be true whether performance in measured in terms of growth in output, investment or employment. A large proportion of firms expected this expansion to continue throughout the 1980s. However, most companies saw this growth related to factors such as increasing demand rather than to the enterprise-zone incentives themselves. Nevertheless, between a fifth and a quarter of firms within the zones considered that their output, employment and productivity would be lower without on-site incentives. Interestingly, a majority of companies both on and off the zones considered that if anything the zones had had a beneficial effect on local business, local economic development, environmental improvements and public- and private-sector investment.

Planning and property implications

About 90 per cent of development in the zones has been for industrial or warehouse use. Most new units are (predictably) small, with 60 per cent less than 450 m² and a third less than 185 m². The designation of zones has certainly encouraged development, but this has varied from region to region. Some – especially those in the south or those with a wide range of incentives such as regional and European Community assistance – have done much better than others. Corby and Wellingborough fall into this category, while others, for example, Milford Haven (Dyfed) and Tayside, have done less well because of factors such as distance from markets, location within depressed regions, and so on.

In general, it seems that enterprise-zone designations have increased the supply of property. Large sites are generally made available, which are more attractive to developers and investors because of more flexible planning regulations, capital incentives and public-sector infrastructural investment. In fact, the P.A. Consultants' 1987 review of enterprise zones concluded that the capital allowances in the zones had begun to attract private finance capital to the north and north west, since incentives raise the yields to those levels required by the investors. For this reason, much of the incentive provided to schemes in the south (and to retail schemes virtually everywhere) was unnecessary – the projects would have in any case gone ahead.

The development of the zones has had implications for surrounding areas. Capital valuations tend to be much higher for on-zone property. Rents also tend to be higher, particularly for smaller premises, and this differential has, if anything, increased. However, this variation is due largely to the rate relief enjoyed by on-zone occupiers rather than to any obvious collapse of the property market outside of the zones. In effect, landlords have clawed back for themselves an increasing proportion of rate relief on premises within enterprise zones. Additionally, the zones have not created industrial dereliction, and properties vacated by tenants moving to the zones have been re-occupied through normal filtering processes.

The costs and perceptions of incentives

The costs of the experiment arise in a number of ways. Between 1981–2 and 1985–6, total public costs associated with enterprise zones amounted to a gross figure of almost £400 million. Of this total, 21 per cent came through rate relief, 38 per cent through capital allowances, 7 per cent through land acquisition by public agencies and 34 per cent through infrastructural expenditure. This suggests that each of the 200,000 hectares developed by 1986 had required almost £200,000 of public expenditure. On the infrastructural side, most of the investment had come from Urban Programme and derelict-land grant sources. Some expenditure would have occurred irrespective of designation, but almost three-quarters of total spending occurred as a result of enterprise-zone declarations. Of this net figure, about 51 per cent represents capital allowances, 28 per cent rate relief and 21 per cent infrastructure and land-acquisition costs.

Some zones, notably Corby, Tyneside and the Isle of Dogs, have received a much larger proportion of total spending than have some of the later designations, such as Milford Haven, Scunthorpe–Glanford and Middlesborough. The total cost of each additional job is estimated at about £8,500 in the zones, and about £23,000 for additional jobs within the local area as a whole.

Finally, how did companies assess the incentives on offer within enterprise zones? Rate exemption was the most important, particularly for pre-designation companies and new start-ups, but less so for branch plants where the benefits might accrue to head offices. Rate exemption was also seen as the most important factor in facilitating additional investment and in explaining why companies located in the zones. On the other hand, a substantial minority considered that infrastructural investment was a significant factor for location within an enterprise zone. These perceptions are predictable, if somewhat misguided. Most additional public-sector investment has come through capital allowances. Of the potential beneficiaries of enterprise zones – occupiers, developers, landowners and investors – it is the investors who gain the most. Of the incentives on industrial property in the north, perhaps 90 per cent goes to investors.

Enterprise Zones: Conclusions

In some respects, enterprise zones have been relatively successful. Between 1981 and 1986, about 1.5 m² of increased floorspace was constructed, representing a rise of 60 per cent on designation totals. About 13,000 net new jobs have been created. Environmental standards have improved in some of the older zones, and local physical and economic development has been improved. Firms within the zones seem to have been doing better than equivalent companies elsewhere. But there have been costs.

Job-creation costs for the zones and the areas around them are high. There has also been a considerable degree of unnecessary subsidy. In the south and in retail developments especially, many projects would probably have in any case gone ahead. Of central importance to a government committed to reduced public-sector have been the incentives and other aspects of public-sector intervention have proved crucial. The most successful zones have not been where the public sector has withdrawn, but where (as in Corby or the Isle of Dogs) one public-sector agency has been able to acquire and to service land for private developers and occupiers. Whether this reality was important or not, the Secretary of State for the Environment announced in late 1987 that there would be no general extension of the enterprise zones in England (Ridley, 1987). As the next chapter shows, by the late 1980s other mechanisms for restructuring the older industrial areas were of greater appeal to the government.

Freeports

Although enterprise zones have been the most important of the liberalizing initiatives, other developments fall into much the same category. This is perhaps particularly true of Freeports, although whether these can be regarded as central to inner-urban policy is very debatable.

Abroad, the idea of Freeports has been current for many years. In the most common form they can be seen as 'enclaves treated as being outside the customs territory of the host state, where goods might be imported, stored, processed and re-exported without becoming subject to customs duties' (Morison, 1987, p. 103). They have been particularly successful within the Pacific-rim economies, and perhaps as much as one-fifth of world trade went through them by the mid-1980s. However, as Morison makes clear, their relevance to the British economy has always been doubtful because a large proportion of industrial imports enter free of duty, and membership of the European Community has limited their scope.

Nevertheless, the government in 1983 announced the creation of Freeports. Although it was made clear that their incentives would simply extend to exemption from basic customs control (unless or until their products entered European Community territory) and that there would be no other special assistance, the government received 45 applications for designation. Six Freeports were declared, in Southampton, Liverpool, Birmingham, Cardiff, Prestwick and Belfast. Interestingly enough, as with so many spatially-discrete policies, Wales, Scotland and Northern Ireland did not miss out. It is doubtful whether this was wise. Some of the more obvious locations, such as Felixstowe (Suffolk) were not granted Freeport status. Instead, the initiative became a mechanism through which the government hoped some form of economic activity would occur in what were, generally, depressed economies. Progress was very slow in the mid-1980s, except in Liverpool, where the Docks and Harbour Company actively developed and promoted the Freeport. Elsewhere, Freeport managers (generally private-sector companies) have presided over enhanced property development. Since many of the concessions within Freeports can be obtained anywhere through clawbacks and concessions, it is hard to see what activity will be forthcoming, other than would have occurred in any case.

Simplified Planning Zones

Finally, we should briefly mention Simplified Planning Zones (SPZs). In 1987, the government announced that local authorities would be able to declare SPZs under legislation contained in the Housing and Planning Act 1986 (Waldegrave, 1987). SPZs are based on the liberalized planning regimes pioneered within enterprise zones. They allow local authorities to give advanced planning permission for specified types of development within defined areas. Individual applications for development will not be required if the proposal is compatible with overall regulations.

The approach was regarded by the government as being of particular benefit to the older urban areas. However, it is difficult to see why SPZs should make any difference. Most applications for development are dealt with very rapidly by urban local-planning authorities, which are mostly eager to acquire new economic activity. In any case, if enterprise zones are the model, the overwhelming evidence is that financial incentives have proved the driving force, and that liberalized planning regimes have been of limited significance.

Liberalization and the Cities: Conclusions

Three brief comments might usefully be made here. First, the rhetoric of deregulation has not matched the reality: enterprise zones and Freeports have largely proved mundane. The worries expressed by many in the cities and elsewhere about the possible return of social and environmental conditions reminiscent of Victorian capitalism have proved groundless. Second, the major initiatives discussed here seem to have been poorly designed. For example, if enterprise zones were supposed to create jobs or to boost new enterprises or to engender 'high-tech' industries, a more precise targeting of incentives could have been devised instead of the blanket concessions on offer. Third, to re-iterate a theme central to this chapter, the private sector does not expand as the public sector withdraws. The evidence from this, and from the next chapter, suggests that market investment follows State expenditure.

6
Conservative Inner-Urban Policy after 1979: Urban Development

The two previous chapters explore aspects of inner-urban policy after 1979: Chapter 4 examines questions of co-ordination and Chapter 5 focuses on deregulation and enterprise. However, one other theme – development – has proved increasingly dominant in the 1980s. It should be restated that these three aspects of policy, co-ordination, deregulation and development, cannot always be neatly separated – there are grey areas. For instance, a great deal of development has occurred within enterprise zones, and the Urban Programme has presided over a wide range of projects. Nevertheless, this threefold distinction highlights key assumptions underlying different initiatives. And whereas some initiatives have been primarily intended to integrate urban intervention and others to test the efficacy of deregulation, the mid-to-late 1980s saw a marked proliferation in inner-city programmes designed to boost development.

The objective of developing the cities, in both an economic and physical sense, has always been apparent within inner-urban policy. The 1977 white paper, *Policy for the Inner Cities* (HMSO, 1977), proposed a number of development programmes including the creation of industrial sites and units within the older urban cores, and extensive refurbishment of the housing stock. Other programmes outlined in the 1977 statement were less directly concerned with development. Partnerships were to be developed between central and local government, social problems were to be relieved and a better demographic balance was to be achieved between the older cities and their surrounding conurbations.

It is tempting to contrast the 1977 approach with that outlined in the 1988 *Action for Cities* (HMSO, 1988a). There it is made clear that Urban Development Corporations, the spearhead of the development thrust, 'are the most important attack ever made on urban decay' (*ibid*. p. 12). Of the twelve initiatives announced in the 1988 statement, eight can be seen to lie within the general sphere of development. In dealing with questions of urban development, there is some point in making a twofold division between Urban Development Corporations (UDCs) on the one hand, and other development initiatives on the other. UDCs are the most controversial of the urban innovations, and are given separate consideration later in this chapter. Before that, however, consideration is given to other, relatively less important, programmes.

Urban Development Initiatives

Under this general heading we can identify perhaps seven composite programmes, which in some way are designed to enhance urban physical and/or economic development. Some innovations undeniably lie within this definition, including urban-funding schemes, land registers, industry and enterprise, garden festivals and some specific projects outlined in the 1988 *Action for Cities* statement. Two other approaches, somewhat more debatably, lie here as well: training and education, and housing programmes.

Urban Funding

In 1982, the Secretary of State for the Environment, Michael Heseltine, announced the creation of Urban Development Grants (UDGs). These were based to some extent on the American experience of urban development action grants. The American grants and UDGs were based on the concept of leverage, by which minimal public-sector support would stimulate substantially greater private-sector investment. UDGs were thus intended as last-resort grants, which would push otherwise unprofitable projects into viability. No restriction was placed on schemes supported by UDGs – industrial, commercial, residential, recreational or community-based projects or combinations of these could all be assisted. However, after 1984 a greater emphasis was placed on schemes designed to boost the physical regeneration of the inner cities. The private-sector contribution to these schemes was intended to be several times larger than that of the public sector.

A major review of UDGs was published in 1988 (Aston University, 1988). In the same year, members of the consultancy team undertaking this evaluation also outlined the major findings (Johnson, 1988; Pearce, 1988). By 1986 UDGs worth £78 million had been approved for 177 projects, with a public-to-private-sector ratio of 1:4.5. About £350 million will have been invested by the private sector when the approved projects are completed. Of the £78 million worth of UDGs, £35 million was for commercial developments, £26 million was for industrial developments and £17.5 million was for housing developments.

In some respects, UDGs have performed reasonably well: very few projects implemented through UDG have failed and most provided tangible benefits. A large proportion had substantial additionality, although some projects could have been supported by other than public-sector funds, bearing in mind that UDG is supposed to be a last-resort lender. In addition, about 20 per cent of projects, mostly larger than average, had little or no additionality; that is, they could have been implemented by the private sector alone.

UDGs also helped to fund the construction of about 2,700 dwellings, at an average grant of £7,800 per dwelling. Many of these houses represented virtually the only new private-sector accommodation available in the inner cities. About three-quarters of land developed with the aid of UDG had previously been under-used, vacant or derelict. About half the developers associated with UDG-supported projects intended to further their urban activities. Finally, UDGs helped to increase otherwise static urban-land prices, and many businesses moving into or owning premises supported by UDGs intended to boost their output.

However, it was also clear by the mid-to-late 1980s that there were drawbacks with UDG. Many urban areas had not really benefited to any great extent. Much of the investment was directed to Greater London, the West Midlands and the North West. Some authorities were better organized in relation to UDG submissions, and some areas were simply not regarded as attractive to the market – even when UDG was taken into account. There was a lack of good, implementable schemes that highlighted weaknesses in terms of development and appraisal skills in both the public and private sectors. The effect was a substantial underspending of UDG. In the three years between 1983 and 1986, UDG allocations of £148 million was matched by ultimate expenditure of less than £50 million.

There were other problems in the UDG programme. First, the extent and costs of job generation were disappointing. It is difficult to assess the total effect on employment of any assistance, but the 1988 *Action for Cities* statement (HMSO, 1988a) argued that UDGs provided nearly 30,000 jobs. It was originally assumed that UDG could create jobs at around £5,000 per post. However, more detailed analysis based on an evaluation of a sample of companies within UDG schemes were less sanguine (Aston University, 1988; Johnson, 1988). Perhaps only about 30 per cent of jobs within UDG-supported schemes were new to the area, although of these, three-quarters were new to the national economy. Approximately 20 per cent of employees were previously unemployed. In terms of the cost of job generation, when taking into account only those jobs created by public-sector support, and that are new to the economy, total costs are two or three times greater than originally assumed. A second major defect of UDG was that it did not make much difference to cities as a whole. It had a pepper-pot effect that did not represent a comprehensive approach to urban regeneration.

Partly to overcome these constraints, Urban Regeneration Grants were announced in 1987. These by-passed local government and allowed developers who intended to regenerate areas of 20 acres or more to approach the Department of the Environment directly. However, only five were approved before a new City Grant was announced in *Action for Cities* (HMSO, 1988a). The City Grant is to replace UDG, Urban Regeneration Grant and private-sector derelict land grant. The new grant will be paid by central government, thus abolishing the 25 per cent contribution that local authorities have made to UDG. The City Grant will be given to private-sector development projects that would not otherwise be implemented, and that can benefit run-down inner-city areas at reasonable cost. Priority is to be given to the 57 authorities submitting Inner Area Programmes.

Land Registers

The problems of vacant and derelict land have been mentioned in Chapter 2. The whole area remains complex. Demand for inner-city land became very depressed in the late 1970s and early 1980s as traditional urban-based function, such as manufacturing and retailing, contracted and/or decentralized. At the same time, supply constraints became more apparent. Public- and private-sector users were both reluctant to put vacant land on the market. Many owners – probably correctly – believed that prices would ultimately rise again, and few wanted to sell land at dramatically lower prices than had been initially paid to acquire it. Public-sector land owners were often under an obligation to obtain the best price possible, which was unlikely to happen in times of such severely-restrained demand and when an urban-land market hardly existed at all. Further, many public agencies wanted land for future development proposals, but a lack of activity within inner-city land markets, and accusations of land hoarding by the public sector, triggered a response from the 1979 Conservative government.

It could be argued that local authorities, in particular, were caught in an intolerable position. In the 1970s they had been required to obtain land for future projects, and they could hardly have been expected to sell at such severely-depressed prices. In many cities, most vacant land was not owned by local government but by the private sector and statutory authorities. Despite these considerations, in 1981 the Department of the Environment announced the creation of land registers.

Land registers record disused public-sector land: between 1981 and 1988, about 50,000 acres were identified and subsequently developed. However, by 1988 the registers still listed 90,000 acres of disused land. In *Action for Cities* (HMSO, 1988a), the government announced that developers wishing to use recorded land could ask the Secretary of State to force its sale, and they could request additional listings. An increasing proportion of land developed after registering was subject to Derelict Land Grant, of which £25 million was, by 1988–9, available specifically to the inner cities.

Land registers have been useful in identifying vacant urban land, and in imposing a degree of dynamism within urban-land markets. However, their real impact should not be over-estimated. Vacant urban land has much more to do with the changing fortunes of the market and declining public-sector resources than with local-government inefficiency. This point was emphasized in late 1988, when a National Audit Office review of Derelict Land Grant concluded that growth in derelictions was outstripping efforts to bring such land back into use with or without grant (National Audit Office, 1988a).

Industry and Enterprise

In the late 1980s, a number of industrial initiatives were announced that were intended specifically to assist the older urban cores (DoE and DE, 1987; DTI, 1988; HMSO, 1988a). These included inner-city offices for the Small Firms Service of the Department of Trade and Industry; grants for inner-city Local Enterprise Agencies (see Chapter 8); higher loan guarantees for small business in the Task Force (see Chapter 4); and investment of about £10 million in managed workshops by English Estates – a development agency of the Department of Trade and Industry. In addition, Urban Programme authorities will also benefit from Trade and Industry incentives: in such areas, two-thirds of marketing, design, management and manufacturing consultancy costs will be paid. In inner cities, within the Development Areas (where regional incentives apply), small companies with fewer than 25 employees will be eligible for 15 per cent Regional Enterprise Grant on fixed assets up to a threshold of £15,000 and a 50 per cent innovation grant up to a maximum of £25,000. Ethnic Minority Business Initiatives are also available, providing venture capital, resources and counselling services for ethnic-minority businesses.

Garden Festivals

The idea of garden festivals was imported from Europe, particularly West Germany, where the concept had been in operation since 1945. The first British festival was held in Liverpool in 1984, the second in Stoke-on-Trent in 1986 and the third in Glasgow in 1988. Additional festivals are planned for Gateshead and Ebbw Vale (Gwent). The idea is that an area consisting largely of derelict and/or vacant land should be landscaped and used for a variety of horticultural, leisure and recreational purposes, normally for a six-months period. The site will attract visitors during the festival, and will later act as a focus for inward investment. In addition, festivals can ensure a concentrated attack on a particular site, can stimulate local employment and help to solve local planning problems (DoE, 1983; O'Connell, 1986).

The festivals have been shown to be relatively successful by some indicators. Between two and four million people visited each of the first three festivals, and the sites were dramatically improved – often at relatively limited net cost. Glasgow, for example, cost about £18 million in 1988 prices, but the site is to be used for a variety of commercial, recreational and residential developments. Some problems emerged in Liverpool when one company associated with the site went into liquidation. However, at relatively limited cost, garden festivals have helped to renew derelict sites and to act as a focus for development.

Action for Cities, 1988

A number of development initiatives were outlined in the 1988 statement, *Action for Cities* (HMSO, 1988a). These include environmental improvement projects; the creation of cultural infrastructure such as the Merseyside Maritime Museum, the Tate Gallery of the North and Granada TV's News Centre in Liverpool's Albert Dock; urban-road projects, such as improving the A13 in east London and the building of a new Black Country spinal road; and the formation of British Urban Development (BUD), backed by eleven construction and engineering companies that will realize profits from management fees and the sale of developed sites.

Training and Education

Inner-city policy has, on the whole, not related to questions of labour supply. However, by 1988 the city technology colleges had been designated in a number of major cities. Additionally, *Action for Cities* outlined in 1988 twelve schools/industry compacts supported by the MSC, in which companies offer firm job prospects to pupils achieving certain standards. In training, *Action for Jobs* programmes involved almost half-a-million inner-city residents by 1987. New projects outlined in the 1988 *Action for Cities* statement include establishing urban information points, decentralizing training services and creating specialist staff for those with particular problems.

Whether these types of initiative ought to be included within the development theme is arguable, but labour-supply considerations may be seen as constraining urban output. Enhanced educational and training standards will have demand implications for services in such areas as tourism, leisure and recreation. Better-trained and more-qualified workforces also correlate positively with the creation of small companies.

Housing Programmes

Housing has not figured prominently in inner-urban policy. Some Urban Programme resources have been used for housing rehabilitation, the London Docklands Development Corporation has presided over substantial housing developments (see below) and, as discussed above, UDG has boosted inner-urban residential schemes. But housing has tended to be subsumed within other programmes. However, by the mid-to-late 1980s, this was beginning to change, with additional resources being allocated for the management, repair, refurbishment and restructuring of housing. The Priority Estates Project, designed to improve the management of public-sector housing, is discussed in Chapter 4. Other schemes include the Estates Action programme, which, with resources of over £140 million by 1988–9, is designed to improve the management of council accommodation, to transfer dwellings to local management trusts, to attract in private-sector resources and, through Community Refurbishment Schemes, to use local people to improve their own accommodation.

More importantly, the Housing Act 1988 proposes to create Housing Action Trusts (HATs) that would be centrally-appointed bodies with planned resources of £125 million in their first three years of existence. Their functions would include taking over and improving public-sector estates and passing them onto new owners, such as private landlords, housing associations and tenant co-ops. HATs clearly reflect the Conservative government's view that urban development requires the creation of autonomous, centrally-controlled and funded agencies, of which the best example is the Urban Development Corporation.

Urban Development Corporations and their Alternatives

In England, the single most important device through which urban development has been channelled has undoubtedly been the Urban Development Corporations (UDCs). However, the real significance of the corporations extends beyond this. The corporations have been the most intensive and the most controversial inner-urban English initiative ever devised. The policies and attitudes of the corporations have, moreover, been thrown into sharp focus by the emergence of alternative approaches to wide-scale urban regeneration, which have been promoted elsewhere in the UK, notably in Scotland. In the remainder of this chapter we examine the evolution and successes of UDCs, and evaluate these within the context of contrasting (but comparable) experience.

UDCs: Origins, Evolution and Powers

The concept of an independent development agency has a long tradition in Britain. From 1946, for example, New Town Development Corporations were given extensive powers to acquire and develop land as part of the national new-town programme. The success of these development corporations prompted organizations such as the Town and Country Planning Association in the 1970s to suggest that some kind of development agency for the inner cities should be created (Town and Country Planning Association, 1979). These sentiments appealed to the Conservative government elected in the late 1970s. Under the aegis of the Secretary of State for the Environment, Michael Heseltine, Urban Development Corporations were given legal definition in Part XVI of the Local Government, Planning and Land Act 1980.

UDCs are primarily intended to secure the regeneration of their designated areas by bringing land and buildings into effective use; by encouraging the development of existing and new industry and commerce; by creating an attractive environment; and by ensuring that housing and social facilities are available to encourage people to live and work in the area. However, it was always intended that UDCs would act primarily as catalysts, providing a framework within which the prime agency for urban renewal – the private sector – would thrive (DoE, 1980). UDCs were to concentrate on the reclamation and servicing of land and not, on the whole, on the development of that land.

UDCs were given certain powers to achieve these objectives. These include the ability to acquire land at values that take no account of development gain arising from their designation and activities. Land was vested in UDCs through statutory orders from the Secretary of State for the Environment. The corporations are also the development-control authority for the designated area, so they may thus determine planning applications made to them or deem themselves planning permission by resolution. They are not, however, plan-making authorities: these responsibilities remain with local authorities for the area. UDCs must take account of such plans in formulating their own development proposals but are not bound by them. It is important to stress that UDCs are not the lead agencies in such services as housing, education and health, which remain with local authorities and other public agencies. It is therefore important for UDCs to attain appropriate working relations with the relevant public bodies. This has not always happened, as is examined later.

Control of UDCs lies with the Department of the Environment. The department issues financial memoranda and written guidance, and discusses with each UDC the form and content of the annual corporate plans. Once approved, these are used as the basis for governmental policy review. The department may be involved separately in approving particularly large schemes or where projects involve UDCs disposing of land at less than market values. The department is also centrally involved in selecting the chairpersons and boards of the UDCs. Key figures in the early evolution of the corporations included Nigel Broackes, Chairman of the London Docklands Development Corporation (LDDC), and Basil Bean, Chief Executive of the Merseyside Development Corporation (MDC), who were well known within City and market institutions. Not surprisingly, under this kind of control the UDCs have tended to pursue market-orientated strategies within their designated areas (Ward, 1987).

UDCs have been designated in two distinct phases. In 1981, the LDDC and the MDC were created, and much of the debate surrounding UDCs has inevitably centred on these institutions. In 1986, the Secretary of State for the Environment, Nicholas Ridley, announced the creation of four more UDCs (see Figure 2, p. xii), each expected to spend about £150 million in the first five years of their existence. These were to be on 2,400 hectares in the Black Country, 4,500 hectares on Teesside, a Tyne & Wear Corporation along the River Tyne and one in Trafford Park (Grt Manchester). In 1986, the Secretary of State for Wales announced the creation of a Cardiff Bay Corporation. This programme envisaged a barrage across the bay, the building of over 4,000 dwellings and the creation of 20,000 jobs. In the following year, three small UDCs were also designated, each intended to spend about £15 million over four to five years. These were located near Temple Meads station in Bristol, along the Aire-Calder navigation and south of the city centre in Leeds and in central Manchester. Finally, in 1988 in *Action for Cities*, it was announced that the Lower Don Valley in Sheffield was to be a UDC, covering about 2,000 hectares with a seven-year budget of £50 million. At the same time, the MDC was doubled in size by the addition of 800 acres on either side of the River Mersey in which about £90 million will be spent over eight years. At the launching of *Action for Cities*, it was announced that no more UDCs were planned for the foreseeable future.

The Merseyside Development Corporation and the London Docklands Development Corporation

Much of the debate surrounding UDCs has concentrated on the specific evolution of the two longer-standing corporations established in the Docklands of Liverpool and London. The remaining sections of this chapter explore some of the crucial issues surrounding these two initiatives.

The MDC, as established in 1981, consisted of three separate sites: one near Liverpool city centre, one at Bootle and a third on the Wirral Peninsula. The boundaries were tightly drawn by the Department of the Environment in order to avoid encroachment onto port land or on sites subject to local-authority initiatives (National Audit Office, 1988b). However, most informed commentary suggested that its 860 acres had always been too small, and in 1988 it was virtually doubled in size.

The MDC was intended to regenerate one of the most depressed parts of a singularly-depressed city region. Most of the land and buildings within the designated site were vacant, derelict and even polluted. About only 400 people lived within the corporation's boundaries, and only about 1,500 worked in the area. The MDC had also to operate in what it saw as a particularly unfortunate political environment. The marked radicalization of Liverpool City Council in the early 1980s was clearly unlikely to assist the corporation in its demand-led, market-orientated strategy, which was based on public-sector pump-priming resources levering out greater private-sector investment (Adcock, 1984). Nevertheless, in the early-to-mid 1980s, the MDC pursued a strategy based on the physical restoration of land and buildings, infrastructural investment, economic regeneration and social renewal. However, by 1986 it had become evident that the collapse in manufacturing had been so severe that alternative approaches based on tourism and retailing should be given much greater prominence (MDC, 1986).

The MDC has had its successes (Wray, 1987; National Audit Office, 1988b). By 1987 it had, for example, reclaimed more than 240 acres of land for commercial or housing development. It had invested more than £30 million into the refurbishment of the Grade-I-listed Albert Docks. It had brought other land and dwellings into effective use, partly because most leases for commercial use were relatively short and freehold interests could be rapidly acquired. In addition, the corporation organized the 1984 International Garden Festival on a 125-acre reclaimed site, and it has helped to boost new enterprises through managed workshops.

On the other hand, there have been problems, such that in 1988 the National Audit Office could argue (pp. 4–5) that 'there was an urgent need for a fundamental review of the MDC, including consideration of its management, development strategy, the size of its area and the timescale and resources needed to establish the Corporation's effectiveness'. The MDC cannot be held responsible for some problems. Its limited size from 1981 to 1988 has already been mentioned. There is also the constant difficulty that market-led strategies need private-sector investment. By the end of 1986–7, MDC expenditure of about £140 million had resulted in about only £20 million of private-sector investment in commercial and industrial developments, although at least another £40 million was committed.

To some degree, the MDC has been placed in a particularly onerous position because of its having to compete for industrial and commercial investment with surrounding new towns and the local enterprise zone and Freeport. Nevertheless, the National Audit Office concluded that much of its marketing 'has lacked focus and drive and has been limited in scope' (*ibid*. p. 4). Other operational problems and constraints have affected the MDC. There has been little development at Bootle and on the Wirral site. Between 1981 and 1987, the number of permanent jobs had increased by only 1,000, although training has been substantially boosted by the corporation. Of the 1,400 dwellings originally planned in 1981, work had started on only 170 by the end of 1986–7. The garden-festival site proved a constant headache. The MDC incurred capital and operational costs in keeping the now renamed Festival Gardens open. In 1986 it entered into an agreement with a private firm to operate the gardens, despite clear warnings from a number of sources about the viability of the company concerned. It subsequently went into receivership in that same year, with estimated debts of more than £5 million. Three years of long and protracted negotiations characterized the proposals for an indoor-events arena. In short, the MDC has presided over some successes, but not over what might be termed the regeneration of Liverpool's docks.

This has not been the case with the LDDC, where a very extensive regenerative programme has characterized the corporation's activities in an area of about 5,000 acres downstream of Tower Bridge (see Figure 3, p. xiii). This area had been the subject of a number of plans and initiatives since 1973, when it became evident that the docks and dock-related activities were set to close and/or move downstream to Tilbury and elsewhere. Whatever the causes of this rapid decline in port activities – containerization, competition, labour disputes, poor management or manning practices – by 1981 the docks had to all intents and purposes closed.

Throughout the 1970s, central and local government were responsible for the publication of a range of plans and reports designed to re-invigorate an area whose economic base might have disappeared, but one with tremendous opportunities and where almost 50,000 people still lived (Ambrose, 1986). In 1973, for example, Travers Morgan published a series of development scenarios, and in

1976 a London Docklands Strategic Plan was agreed by the Greater London Council and the five boroughs concerned. This plan argued that the renewal of the area should be based on the twin policies of the construction of social housing and industrial development, and it was to form the basis, in the late 1970s, of the Partnership established between the Labour government and the local authorities concerned (see Chapter 3).

However, the incoming Conservative government of 1979 opted for a different approach. A UDC was to be created – the LDDC – that was to concentrate on attracting in private-sector investment, partly by attempting to overcome what were seen as major obstacles to regeneration: poor communications and infrastructure, a deteriorating physical environment, problems with land assembly and the area's poor image. To the LDDC it seemed apparent that any effective programme for Docklands would need to concentrate on a number of key objectives. These included land acquisition and development, widened housing and employment opportunities, enhanced community facilities and an improvement in the area's image.

These kinds of objectives were to be attained by public-sector pump-priming investment designed to lever out private resources. Between 1981 and the end of 1986–7, the LDDC had spent about £380 million, most of this in the form of grant aid from the Department of the Environment. These resources had encouraged private-sector spending in excess of £2 billion. These rates are likely to be at least matched in the 1990s when, between 1987–8 and 1992–3, the corporation plans to invest at least £600 million – half through land sales and half through central grant (LDDC, 1987).

The redevelopment of Docklands has been achieved through five major policy issues; land reclamation, infrastructural investment, housing, economic development, and employment and social infrastructure. In land acquisition and development, by the end of 1986–7, the LDDC had acquired almost 2,000 acres of land or enclosed water. It had released more than 700 acres of land for development, which amounted to about 30 per cent of the total of derelict land within its designated boundaries (National Audit Office, 1988b). The corporation sells land by tender or by private negotiation. To ensure that developments are completed in time, the LDDC grants building licences to developers but retains the title to the site. In housing, the LDDC grants freehold title to purchasers of individual dwellings. In the case of commercial or industrial development, the corporation normally grants a 200-year lease at a peppercorn rent, without rent reviews. As the National Audit Office (1988b) points out, however, this approach does not allow the LDDC to prevent the intensification of use by leaseholders, nor for the corporation to benefit financially from such developments.

The LDDC has presided over substantial infrastructural investment. For example, the Docklands Light Railway has improved public transport within a traditionally isolated part of London. The initial £77 million budget was split between the LDDC and London Regional Transport, and it involved the building of a railway from Tower Hill in the west through the Isle of Dogs to link up with

Stratford in the north. Later plans involved an additional expenditure of about £150 million to take the railway into Bank, about half of which would be met by the Canary Wharf developers (see below). Plans are also in hand to extend the railway to Beckton in the far east.

In addition, the LDDC plans to spend over £200 million by the mid-1990s on road improvements with Docklands, and the Department of Transport is to spend more than double that figure on linking the area to the national road network. These projects include an improvement to the A13 – the main east-west road – and a new east-London river crossing. Finally, in infrastructure, the LDDC has promoted the new London City Airport in the Royal Docks, with flights to a number of European cities.

There has been considerable development in housing since the creation of the LDDC in 1981. At that time there were 14,000 dwellings within the designated area, only 4 per cent of which were owner-occupied. The LDDC decided at an early stage that this figure should rise to 30 per cent or so, as a result of a house-building programme that was originally set at 13,000 houses within ten to fifteen years. As almost 12,000 houses had been started by 1987, the overall target was raised to 25,000 new dwellings.

The substantial degree of housing investment within the LDDC area has been paralleled by a similar scale of economic development. By the end of 1986–7, some 2.5 million square feet of commercial space had been completed and another 7.5 million was under construction (National Audit Office, 1988b). In addition, at the time of writing it would appear that two other major projects are to be implemented. In the enterprise zone on the Isle of Dogs, the Canary Wharf commercial project – costing between £3 billion and £4 billion private-sector investment coupled with some £250 million public-sector infrastructural costs, and providing over 5 million square feet of office accommodation – is to go ahead. In the Royal Docks in the eastern segment of the LDDC, a comprehensive renewal project is likely to be effected, costing about £2.5 billion. This will incorporate retail developments, hotels, houses and half-a-million square feet of commercial development.

The LDDC sees these kinds of developments as likely to boost jobs within its area from the 1981 figure of 27,000 to around 80,000 by 1991. In an effort to increase job opportunities for the local population, the corporation intends to spend in the late 1980s and early 1990s some £20 million on education and training. Finally, the corporation has become more interested in investing in social infrastructure. In the six years after 1986–7, the LDDC is intending to spend around £100 million on areas such as education, health, community and recreational facilities, which represents a sevenfold increase over its social budget in 1981–6. This enhanced investment will be undertaken in co-operation with statutory educational and health authorities.

Urban Development Corporations: the Debates

The previous section outlines some of the key characteristics of UDCs in general, and the MDC and LDDC in particular. There are, however, a series of debates surrounding UDCs that need to be mentioned. Many of the issues can be categorized into three central questions: democracy and UDCs, their effectiveness and their sectoral implications.

UDCs *and democracy*

One of the major debates surrounding UDCs is the question of democracy. The corporations are imposed by central government and are given powers of land and development that have been seen as 'staggering' (Cullingworth, 1985, p. 282). The Secretary of State for the Environment also selects board members, chairs and deputy chairs. This kind of administrative structure may be regarded as likely to lead to the implementation of policies that are at variance with those favoured by elected local councils. This has been so in the MDC, although efforts were made to foster links with the now-defunct metropolitan county (Boaden, 1982). In any case, development within much of the MDC was relatively modest in the early-to-mid 1980s. The LDDC's position has been rather different.

At a very early stage in the evolution of the UDC concept, major political leaders in east London pointed out their essentially undemocratic nature. A UDC would undermine the ability of the elected local councils to implement the 1976 Docklands Strategic Plan, which had been agreed only after many years of discussion and debate (Leaders of the London Boroughs of Greenwich, Lewisham, Newham, Southwark and Tower Hamlets, 1979). At least to some observers – notably the Docklands Consultative Committee (1985; 1988), a grouping of local political and community interests – these and similar misgivings have proved all too real. The LDDC has not given local authorities long enough to respond to planning applications; some community groups have not been consulted; until 1986, all meetings were held in secret; and the LDDC has paid only lip-service to local plans produced by the boroughs concerned. Interestingly, in its 1988 review of UDCs, the National Audit Office commented on the lack of co-operation between the LDDC and other public agencies, particularly education and health bodies (1988b). The LDDC has not been especially willing to enter in co-operative, collaborative arrangements with existing statutory organizations.

It could be argued that this lack of co-operation between UDCs (notably the LDDC) and local government is predictable. It might be suggested that UDCs can cut through local bureaucracy and need to operate at different time-scales from those prevalent within local government. Market-led regeneration strategies may, in any case, necessitate rather more of a developmental and rather less of a political stance: there is the view that the renewal of London's Docklands raised issues of national significance that could not be left solely to the boroughs concerned.

Whatever the merits of these arguments, there can be little doubt that for much of its existence the LDDC has not sought out partnership arrangements with elected local bodies – it has pursued market-led strategies whose impact on local communities has proved far from advantageous; it has largely ignored local planning procedures; and there has been little public accountability of its policies and spending. The debate should not, however, be left there. The question of democracy has been seen by the LDDC and the second generation of UDCs as an increasingly important issue. The LDDC has opened up committees to the public. It is to spend more resources on social and community facilities and to employ community liaison officers. In 1987 it signed a memorandum of agreement with Newham regarding the redevelopment of the Royal Docks that stipulated that regeneration would include the construction of social housing, the provision of training and community facilities, and so on (London Borough of Newham, 1987).

Later generations of UDCs also show that agreement between corporations and local government (covering issues such as accountability, social audits, a sharing of planning functions, and so on) is possible, as events in 1987–8 in both Cardiff and Sheffield showed. The issue is not so much that the LDDC was unwilling to effect democratic procedures with elected authorities in the early to mid-1980s, but rather (as is discussed below) that the policies proposed by the LDDC were different from those agreed by local government, and that co-operation would simply have proved impossible.

Are UDCs effective?

UDCs, and perhaps the LDDC in particular, have tended to receive favourable evaluations from many sections of the media and from developers. *The Financial Times*, for example (1986), has suggested that 'there is little doubt that in commercial terms the LDDC has been a success'. Similarly, the *Estates Times* (1987) has suggested the LDDC has promoted substantial change in a development desert. There can be little doubt that, in terms of property development, the LDDC has indeed presided over a dramatic transformation in London's Docklands. However, a number of comments and qualifications ought to be made.

An initial point is that the LDDC could not have been operating in a more favourable environment. The generally massive expansion in financial services, the 'Big Bang', the LDDC's location close to the City, £400 million of public-sector support in the first five years of existence and its extensive powers to acquire land, all placed it in an exceptionally favourable situation. Given these kinds of powers, benefits and resources, it is interesting to speculate upon what the appropriate local authorities might have done during this period.

But it is naïve to assume that the LDDC has always operated as effectively as some reports would suggest. It has not been prepared to consider the wider implications of the massive developments planned for Canary Wharf and the Royal Docks, which will have all kinds of consequences for London and the South East. It has in general avoided strategic issues that would help to identify needs and to assist in co-operation with local authorities (National Audit Office, 1988b). Moreover, it is far from clear that the LDDC has managed to obtain anything like the best financial deals in its negotiations with developers: land appears to have been sold at substantially less than might have been achieved, and as the freeholder the LDDC has retained little control of, nor gained any benefit from, subsequent development.

The question of transport planning within Docklands has also proved chaotic. The LDDC is not entirely to blame for this. However, its activities in relation to the Light Railway, which are documented elsewhere (Docklands Consultative Committee, 1988), reveal serious shortcomings. In particular, it should have been apparent from an early stage that the railway would prove totally inadequate to deal with the demand generated in the Isle of Dogs, especially by the Canary Wharf project. In short, if the LDDC is seen to epitomize the UDCs, then there are questions to be asked about the effectiveness of this form of intervention.

UDCs: gainers and losers

One of the most contentious issues surrounding UDCs is that of gainers and losers. Developing Ambrose's (1986) excellent analysis of this problem, we can identify four major interest groups or agencies whose status has been substantially altered as a result of UDCs, and in particular the LDDC: central government, local government, developers and local residents.

Central governments elected in 1979, 1983 and 1987 have clearly benefited from UDCs in general and especially the LDDC. The government's development thrust has been furthered: a great deal of construction has taken place to the benefit of developers and major building companies. Local councils – most of which are usually Labour controlled – have effectively been by-passed. In London's East End, the social composition of the area will ultimately be substantially changed as a result of the building of 25,000 dwellings – the vast majority of which are for owner-occupation. This will have clear implications for voting patterns at local and national elections.

If central governments have benefited from UDC activities, government on the whole has not. In London, the 1976 Strategic Plan was overturned, and its emphasis on social housing and industry was abandoned by the LDDC's determination to promote the construction of owner-occupied dwellings and commercial development. Local authorities have lost important development-control powers and have had to release land to UDCs at extremely low prices. The three key boroughs in east London (Newham, Tower Hamlets and Southwark) had to accept 4,400 families as homeless in 1986–7 – a threefold increase on 1981–2. Yet the resources available to these boroughs through Housing Investment Programme allocations fell from around £160 million at the end of the 1970s to about

£60 million in real terms by 1987–8. It could be argued that local authorities will gain immensely when the UDCs are eventually wound up. However, the areas they inherit will have been developed according to policies very different from those supported by the local council, which is after all the elected body. Local government will also require appropriate resources to run the physical and social infrastructure the UDCs will have provided. Will these be forthcoming?

These longer-term implications are unlikely to impinge on the development industry, which has done extremely well out of UDC designations. This is not surprising. Many chairs and board members are from companies with some property interests. The first LDDC chair, for instance, was Nigel Broackes from Trafalgar House and the second, Christopher Benson from MEPC. Land has been released to developers for residential and commercial prices well below what the market would bear (Klausner, 1987). Bearing in mind prices prevailing for completed developments within the LDDC, developers and construction companies have gained enormous profits. Obtaining planning permission for developments has been much easier than would otherwise be the case. Wider social policies, which might have been incorporated into regenerative programmes effected by local government, have been muted if not abandoned. Massive infrastructural investment undertaken by UDCs has made the marketing of developments much easier and development costs lower. In some cases grants have been made available from UDCs for the implementation of specific projects.

The development industry has, in brief, benefited from UDCs. Nowhere is this truer than in the LDDC where, as well as the benefits listed above, additional incentives were available in the Isle of Dogs enterprise zone. Between 1981 and 1987, for instance, public subsidies of £130 million were forthcoming for rate relief and capital allowances within the zone. This is surely unnecessary public investment. The regeneration of the Isle of Dogs would have gone ahead without UDC support.

Finally, what of the local residents? Much of the depressing evidence emanates from the LDDC. In terms of employment, the corporation argues that between 1981 and 1987 about 10,000 jobs were brought into the area. However, during this period local unemployment actually rose from 3,500 to 4,400. This is not surprising. The LDDC has used compulsory-purchase powers to acquire and to remove transport and manufacturing jobs. These accounted for about 60 per cent of local jobs in 1981, but only half that figure by 1987. Indeed, almost two million square feet of industrial space was lost on the Isle of Dogs between 1981 and 1987 – mainly to be replaced by expensive residential development (Docklands Consultative Committee, 1988). Little new industrial development has been forthcoming. Of the new jobs in the area, many are transfers from other parts of London, and have often involved substantial net job loss to the city (Colenutt, 1987). For example, in moving to Wapping, News International cut its workforce from 5,000 to 1,500. The new jobs that do arrive are not usually taken by locals – only 10 per cent of jobs on the Isle of Dogs have gone to locals (Tym and Partners,

1987). Few of the jobs to emerge as a result of the Canary Wharf development will be taken by locals. Their skills are unlikely to dovetail into demands of major financial institutions, except at the catering and cleaning level. Local training is to be improved by the LDDC, but between 1981 and 1987 it spent only £2.2 million on its two major initiatives, neither of which deals with the over twenty-fives (Docklands Consultative Committee, 1988).

Many of the comments outlined above have been reiterated in a pointed House of Commons report (House of Commons Employment Committee, 1988). This report accepted that unemployment had risen within the LDDC from about 3,500 in 1981 to about 5,000 in 1986; that approximately eighty companies had been relocated out of the area through compulsory purchases implemented by the LDDC; and that many locals were not gaining from the corporation's activities. The recommendations of this committee included a suggestion that UDCs should make greater efforts to assist local residents; that local skills should be enhanced; that industrial uses should be retained where possible; and that as far as possible jobs would be directed to local residents. This is far removed from the commercial emphasis of the LDDC in its early development.

Local residents within east London have not benefited from UDC designation in other respects. Most transport investment has gone on road construction and not on public transport. Many of the deliberations of the LDDC have been taken with little or no public scrutiny. Very limited resources were allocated to general social and community facilities in the first six years of the LDDC. Conditions for many minority groups have worsened. For example, little sheltered housing has been constructed for the old and disabled. Physical standards for ethnic minorities have deteriorated, with much higher proportions of Asian families living in overcrowded conditions when compared with white households. The apparent drift from manufacturing and transport within the LDDC will have a particular impact on local women, who are unlikely to obtain worthwhile jobs in financial services.

Finally, one issue merits more detailed comment because of its implications for the entire Docklands community – housing. Housing within the LDDC's area has been an exceptionally emotive issue. On riverside sites, sometimes on land previously used for industry, new or refurbished units have been put on the market at prices in excess of a quarter of a million pounds. Yet families living in overcrowded conditions in Newham, Southwark and Tower Hamlets increased from 16,000 to 25,000 between 1984 and 1988, at which date about half the dwellings in these three boroughs were classified as unsatisfactory (Docklands Consultative Committee, 1988).

In studying housing issues, questions of tenure loom large. There are constraining elements affecting the owner-occupied sector, and those impinging on social housing. About four-fifths of housing constructed in the LDDC has been for the owner-occupied sector. To enable locals to buy housing, the LDDC has operated an 'affordable housing policy'. The LDDC negotiates with developers on a scheme-by-scheme basis to ensure that a proportion of dwellings is placed on the market at around £40,000. Priority for all purchases is given first to local-authority and private tenants in the LDDC area and, second, to local-authority and private tenants in any part of the Dockland boroughs, their children and the children of tenants in the first category. Any dwellings not taken up within a specified time are placed on the open market.

A possible two-thirds of dwellings sold by the end of 1986–7 in the LDDC area fell into the 'affordable housing' category. However, there have been abuses of the system. Tenants have sold on at higher prices, and outsiders have attempted to prove local residence. After 1985, these difficulties caused the LDDC to introduce tighter checks and a clawback provision if resale took place within five years. More importantly, some commentators suggest that the central issue is that most locals still cannot afford owner-occupied housing (Docklands Forum, 1987). Three-quarters of local households had an income of less than £10,000 in 1987. For them, social housing remains a priority.

However, the possibility remains bleak of any substantial increase in investment in the construction and repair of social housing for the LDDC area, and for the adjoining boroughs. In 1981, the corporation argued to a House of Lords Select Committee that the tenure mix within the LDDC would work out at about 50 per cent owner-occupied, 25 per cent shared-equity tenure and 25 per cent housing association. By 1987, only about 15 per cent of housing was within local-authority or housing-association sectors. The LDDC, in its 1987 *Corporate Plan* (LDDC, 1987), argued that, although it was not itself a housing authority, it intended to increase this figure by rehabilitating older public-sector estates through a programme of co-operation with local authorities. However, as has already been mentioned, co-operation is ineffective when local councils find their housing budgets so severely pruned. This is due to the attitudes of the central government that created UDCs. One result is that, in 1985–6, enough land was allocated by the LDDC for the construction of 1,600 public sector-units, although financial constraints meant only 600 could be built. For those in social housing in or close to the LDDC, there is little to suggest any short- to medium-term improvement in housing conditions.

UDCs: Conclusions

The rationale for UDCs is presumably that the scale of urban decline necessitates the creation of independent, centrally-appointed development agencies that are free from the apparent constraints of local government. Seven or eight years into the experiment, this cannot, as yet, be verified. UDCs have (and no doubt will) preside over a great deal of property development. Some new enterprises have been created, infrastructural investment pursued and inner-city housing boosted. However, several important questions need to be asked. To what extent have UDCs promoted the relocation of economic and physical activity that would have occurred in any case? What benefits have local residents received from UDCs? Have they pursued pragmatic, *ad hoc* programmes of action when longer-term strategic planning was required? And would local authorities, given the same resources, have achieved the same results?

Alternative Strategies towards Urban Regeneration

Much of the debate about urban regeneration in England has focused on UDCs. However, alternative approaches towards the comprehensive renewal of older urban areas have been devised elsewhere, both in England and particularly in Wales and Scotland. In England, for example, a number of authorities such as Birmingham have created separate development companies. The structure, powers and resources of these companies vary, but they are all intended to achieve the regeneration of specific areas by combining the talents and attributes of the public and private sectors. They are usually created by local government, but have wide representation from financial and development sectors, and hope to obtain funding through European money, Derelict Land Grant, City Grant, and so on. Central government can sometimes initiate these rather more informal approaches towards economic and physical renewal. For example, in 1988 the Secretary of State for Wales, Peter Walker, announced a £500 million revitalization programme for the valleys of South Wales, which would involve land reclamation, business advice and support, factory construction, housing rehabilitation and infrastructural investment, and which would be implemented by local authorities, the private sector, central government and the Welsh Development Agency. However, in seeking alternative approaches towards urban revitalization than those implemented by UDCs, it is the Welsh Development Agency's Scottish counterpart that offers the most interesting experience.

The Scottish Development Agency (SDA) was created in 1975 by the Labour government to revive the Scottish economy, and to co-ordinate industrial intervention with environmental improvements. In the mid-1970s, it was seen by the Labour administration as part of a comprehensive strategy towards regional economic and physical development in that the SDA's activities were to be mirrored in Wales by the Welsh Development Agency (WDA) and in England by the National Enterprise Board (NEB). However, although the WDA has remained, it has not been as interventionist in terms of urban policy as has the SDA. In England, the NEB – designed to enhance State investment in production, especially in areas of high unemployment – had minimal impact, with only about sixty firms ever supported, and most of these in the south of England (Lawless, 1981b). Despite this limited effect, the anti-collectivist Conservative government elected in 1979 was ultimately to abandon the NEB as an independent force.

In Scotland, events were to prove different. The SDA was given (and has retained) wide powers over both the hard and soft aspects of economic development. Hard aspects include land assembly, property development and management and environmental improvements. The SDA has the additional facility to make loans, provide equity investment and establish a range of business services (Gulliver, 1984). The SDA also operates within a very different political and administrative environment to that in England (Morison, 1987). The Secretary of State for Scotland, operating in effect as a regional political chief, is in an ideal position to co-ordinate and innovate public policy. The types of conflict between different tiers of government – so obvious in England throughout the 1980s – are, while far from unknown, less apparent. Intensive political contact in what is a relatively small country, has tended to soften and shift policy disagreements.

By the mid-1980s, the SDA's activities were primarily directed towards attaining four major objectives (Aitken and Sparks, 1986). Some of these are not of immediate significance to the urban debate. Two of its goals are to expand the range of service-sector jobs and to encourage the creation of small firms. A third objective, that of enhancing the electronics' sector within Scotland, should be discussed in a little more detail because of its apparent success, and because many English cities would wish to expand into this sector.

The SDA's package of assistance to firms in the electronics' sector who wish to move into Scotland from abroad, includes help with premises, business and training advice and investment finance. This assistance, aimed particularly at Japanese and American corporations, has in some respects undeniably been successful. By the mid-1980s, about 45,000 people worked in electronics in Scotland, with a particularly high concentration in and around the new town of Glenrothes (Fife). This employment figure probably represents the highest density of those working in electronics in any European region. However, this approach towards electronics is not without its critics (Moore and Booth, 1986): few jobs have been created for the long-term unemployed; a relatively small proportion of posts are for managerial and technical positions; and the indigenous sector remains small, as do total exports and research and development capacities. A great deal of the investment represents multinational corporate investment by firms wishing to use cheap assembly labour in the European Community.

The most important aspect of SDA policy is that of urban renewal. Boyle (1988) argues that the SDA has evolved its urban interventions in four stages. With some apparent reluctance, in the 1970s, the SDA was persuaded to co-ordinate the Glasgow Eastern Area Renewal project (GEAR). This programme (see Chapter 3) involved the spending of over £400 million by a variety of agencies on the regeneration of east Glasgow, through the creation of new business centres, industrial development schemes, environmental improvements and substantial housing renewal and rehabilitation (Leclerc and Draffan, 1984). The second phase of urban policy effected by the SDA, was its creation of two Task Forces in the late 1980s in Glengarnock (Strathclyde – a steel-closure area) and Clydebank, where the Singer plant closed. In both areas the SDA created programmes designed to enhance industrial property, to help small firms and to reclaim and improve land (Gulliver, 1984). The private sector was to be harnessed to stabilize – if not increase – local employment.

The SDA's third urban regeneration phase, according to Boyle (1988), occurred in the early 1980s. The agency identified a number of locations for area or integrated projects, of which at least seven were undertaken in places such as Motherwell, Leith (Lothian) and Dundee. One important aspect of these projects was that the SDA entered a non-binding agreement with other local bodies, such as the local authorities and the MSC, to define objectives, funding and phasing. The SDA thus co-ordinates an economic and physical strategy that is implemented by existing statutory bodies. Finally, in its most recent phase of operation, the agency has shifted the emphasis towards self-help models, where the SDA supports attempts by local communities (including both the public and private sectors) to initiate and to effect local economic development programmes. As part of this approach, the SDA has supported enterprise trusts designed to provide support for small companies.

Not all evaluations of the SDA's urban policy have been entirely favourable (Booth, Pitt and Money, 1982; Morison, 1987; Boyle, 1988). In the case of GEAR, much of the investment would have occurred and, in some respects (notably job generation), that project was far from successful. Its administrative structure was Byzantine. The Task Forces were reactive responses on the part of the SDA to specific closures. It is interesting to note that no similar exercises were launched when the Scottish vehicle industry – located at Linwood (Strathclyde) and Bathgate (Lothian) – collapsed.

Boyle (1988) also argues that the new emphasis on public–private partnerships seeking to enhance business confidence and investment contains unfortunate implications for more depressed Scottish communities. A move towards supporting business activities is likely to lead to more investment in favourable rather than more depressed economies, as the SDA attempts to assist 'winners'. Issues of poverty and deprivation will become less important. Indeed, the trend towards business support may eventually undermine and replace programmes intended to provide social, physical or economic investment for more depressed localities.

Nevertheless, the SDA's approaches towards urban policy have been innovative in a way that the overwhelmingly property-orientated UDCs have not. The Task Forces and Integrated Projects attempted to co-ordinate a range of relevant public bodies, including local government, into coherent responses towards problems of deprivation and economic decline. The SDA has been prepared to experiment with a variety of policy initiatives in a total urban budget of perhaps £220 million (in historic prices) between 1976 and 1990 (Boyle, 1988). Relatively small sums from the SDA have levered out substantially greater private-sector investment. And Glasgow, at the nadir of urban deprivation in Scotland, has improved relatively and absolutely in recent years, as a result of the activities of the SDA and other bodies. Private-sector investment has increased; some middle-class households have returned to the city; and service-sector employment has increased.

The SDA has had a wider set of powers than those readily available to UDCs in England, but it has used them. For much of 1980s, it has devised urban policies of a more imaginative and consensual nature than central government was able to create in England. The National Enterprise Board might never have come anywhere near what was required, but the principle was right: England needs a development agency or agencies with a wide range of social, economic and physical powers that are able to intervene in more depressed localities through corporate programmes emanating from the full range of public- and private-sector organizations. It is to be hoped that recent changes within the SDA (notably its drift towards self-help, business-orientated projects) will not allow it to lose sight of wider goals in Scotland, where extensive areas of economic decline and outright poverty can still be identified.

7
Inner-Urban Policy: the Role of Local Government

Mention has already been made of local government's role in implementing programmes for inner-urban re-invigoration. The Urban Programme and the Partnerships, for example, utilize local authorities in an obvious and intimate manner. On the other hand, many of the innovations developed by Conservative governments (discussed in Chapters 4–6) employ local councils in a subordinate manner or positively by-pass this tier of government. In many respects this seems unfortunate because many of the more innovative approaches to urban economic development and employment generation have emanated from local administrations. This chapter attempts to remedy this deficiency by analysing local government's role in inner-city policy in a threefold manner: first, local economic development is placed in an historical and institutional context; second, an indication of the range of local-government inner-urban initiatives is presented; and finally, some of the key debates and tensions are explored.

Before embarking upon more detailed analyses of local government's contribution, one or two initial comments ought to be made. Not all local-authority initiatives have been undertaken by councils that have severe inner-city problems. Some innovations have been implemented by local government in association with central government, the private sector or other statutory bodies. Not all local authorities have been particularly interventionist in this area. Variety has characterized the activities of councils that have effected local economic development programmes. For this reason, generalizations are difficult. Nevertheless, it would be impossible to evaluate British inner-urban policy without considering the role and impact of local-government-inspired policies in the areas of economic development and employment generation. These initiatives may have emerged from objectives very different from those that have fuelled central government's inner-city programme, but aims such as fostering output and productivity, improving skills levels, enhancing industrial infrastructure and creating jobs have been at the core of both strands of inner-urban policy.

Local-Government Intervention: the Historical and Institutional Context

Local economic development is not new; the municipalization of local services was very evident in the nineteenth century (Hennock, 1973). The economic decline in the inter-war period encouraged a number of development associations designed, among other things, to attract mobile industry. However, after 1945 there was a marked decline in municipal activities and local economic development programmes in general, as nationalized industries took over the running of many utilities and economic decisions relating to, for example, taxation, demand management and regional policy were assimilated and retained by central governments of both major political persuasions. As Camina showed in the early 1970s, although many county councils and county boroughs employed some form of industrial development officer, most activities were limited to promoting the benefits of the area concerned and to providing land and sometimes industrial premises (Camina, 1974).

In the following decade or so, there was a substantial increase in the range and intensity of economic development initiatives implemented by local government (Chandler and Lawless, 1985; Mills and Young, 1986; Sellgren, 1987). Virtually all the metropolitan districts and most counties were undertaking a number of economic development functions and even smaller district councils were becoming more active. Mills and Young's survey (1986) of local authorities and their economic development policies and staff is especially useful in this context. Some of the more pertinent conclusions from their comprehensive analysis can be mentioned. It appeared, for example, by the mid-1980s that programmes for economic intervention were becoming as well defined in rural and non-metropolitan areas as in the major cities. Intervention was wide ranging and innovative – many councils provided some form of financial assistance to companies through grants and loans. Overall, resources allocated to economic development in 1983–4 amounted to £145 million of capital spending in a survey of 161 administrations, and a total revenue budget (in this case for 177 councils) of £77 million. The authors conclude that in general, local economic development has and will continue to expand – despite a clear shortage of resources and a reluctance on the part of many authorities to devise the appropriate corporate structures local economic development necessitates, or to provide economic development officers with the business skills their posts increasingly require.

Some indication of the range and structure of local economic activities can be obtained from Sellgren's (1987) analysis of the Local Economic Development Information Service (LEDIS), published by the Planning Exchange in Glasgow. This evaluation concludes that local authorities instigated about a third of all local initiatives and that many of these were ultimately established as legal entities separate from the founding agencies. Even then, however, almost two-thirds of local initiatives used local-government finance in some way or other to effect a range of policies, of which providing information and advice to companies, labour training, provision of land and/or premises, advertising and providing finance for local firms were the most important. Most activities were designed to promote the creation and permanence of new, small firms, and were implemented by a range of public- and private-sector bodies – local government being the most significant.

The available evidence indicates a significant increase in the 1980s in both the number of authorities undertaking local economic development functions and the range of initiatives. This increase in local economic intervention has been ascribed to a number of factors (Mawson and Miller, 1986). The 1974 reorganization of local government created larger metropolitan counties that pursued research and policy formulation in areas such as economic development. The Inner Urban Areas Act 1978 (see Chapter 3), gave many urban councils powers to declare industrial improvement areas and to prepare industrial sites and refurbish buildings. The Urban Programme acquired a more explicitly-economic emphasis.

In addition to these top-down pressures, in the late 1970s and early 1980s, it became clear that many Labour-controlled urban authorities began to adopt more radical approaches towards policy areas such as local economic development. Many cities in the older industrialized regions encountered dramatic declines in output and employment. Such cities as Sheffield and Birmingham were losing in excess of 1,000 manufacturing jobs each month. To councillors and officers in some authorities, it became apparent that this scale of economic retrenchment (worsened, as they saw it, by over-tight monetary policy and corporate restructuring tendencies discussed in Chapter 2), required appropriate local political responses. Whereas radical councils had traditionally shown their colours through intervention in aspects of consumption, such as housing or education, a new group of radical, white-collar, professional councillors (brought up on the economic primacy of deprivation highlighted by the Community Development Projects), turned their attention to possibilities inherent in local economic activities.

At the time of writing, an examination of the motives and impetuses behind the increase in local economic innovations after 1979 is of limited relevance: it clearly occurred. In the 1980s, many councils, certainly most local administrations in England, developed some form of local economic programme. This trend raises two issues: first, what kinds of innovations have been developed and, as far as can be ascertained, how successful have they been? Second, what broader debates and conflicts has local-authority economic intervention engendered?

Local-Authority Economic Initiatives

A number of surveys into the range of local economic development initiatives have been undertaken (Cochrane, 1983; Chandler and Lawless, 1985; Cochrane, 1986; Mawson and Miller, 1986; Mills and Young, 1986). These surveys indicate that a great deal of economic development implemented by local authorities has occurred within the following spheres: promotion of the area: the provision of

land and industrial premises; business advice; direct corporate financial assistance; innovation and development; training; municipal activities; and programmes in allied policy areas.

A number of provisos ought first to be made. Some schemes will not fall
conveniently within one of these eight areas. In addition, this list does not reflect
the totality of local-authority intervention in the fields of economic development
and employment creation. Some projects, where local government plays an
important role, such as the Urban Programme, have already been discussed.
Other schemes of a more community-based nature are examined in the next
chapter. One final point, only an indication of developments in each of the eight
main categories can be provided; those interested in exploring the field further
should turn to more detailed references.

Promotion

One of the longest-standing initiatives designed to boost local economic development is promoting the area concerned in an effort to attract in mobile investment.
Authorities may do this in a number of ways. They may advertise nationally
through the Press and television. Trade delegations may be sent abroad to trade
fairs, and so on. Authorities may also organize conventions, to heighten the
awareness of the city or town in question, in financial and development circles.
Enquiries to authorities will usually elicit information relating to a range of
apparent benefits available to any potential investor. Typically, an authority will
refer to local skills, development opportunities, availability of premises and
financial support, local housing markets, leisure and recreational opportunities,
local infrastructure, and so on. Although many authorities persist in publishing
material that often verges on the mundane, there has been a marked improvement
in the standard of publications emanating from some councils (see, for instance,
Sunderland Borough Council, undated).

Although much of the emphasis on promoting the benefits of an authority has
been intended to attract in industrial and or commercial development, there has
been a marked increase in programmes designed to attract tourists and conferences. Conferences have expanded considerably since 1970, and some cities,
such as Sheffield, have tended to use existing infrastructure (for example,
contained in institutions of higher education) to capture national conventions and
conferences. Other councils, of which Brighton and Harrogate are good examples, have implemented or helped to implement the building of major conference
centres that act as magnets for larger, sometimes international events.

While some authorities have attempted to attract more in the way of conferences and conventions to their area, there has been in recent years a much more
significant expansion in activities designed to boost tourist activities. Indeed, it
could well be that tourism will become one of the major areas of economic
development in the 1990s as far as local authorities are concerned.

There are a variety of mechanisms through which local authorities can help boost local tourist activities. They can encourage conservation, thus increasing the appeal of the area to visitors, as has happened with London's Covent Garden and the Piece Hall in Halifax. Under the Local Government Act 1972 and under other legislation, they can provide support for museums, orchestras, theatres and other cultural projects. Bradford, for example, provided financial grant and an appropriate building at a peppercorn rent to house the National Museum of Photography — an outpost of the National Science Museum. In addition, local government can provide appropriate infrastructure, such as car parks, the allocation of land for leisure and tourist activities, package holidays and publicity promoting the attractions of their area, for day and long-term visitors. Although some councils have not been particularly interventionist in the area, others, such as Bradford, Birmingham and Glasgow, have managed to market themselves very successfully as tourist centres.

Promoting an authority to tourists, conference organizers, developers, industrialists and investors, raises a series of potential problems. One is the zero-sum element inherent to many of these initiatives: one authority gains to the detriment of another as investment moves around to find the most attractive incentives. This argument goes further. Firms may move to less economically-active areas, thus relieving inflationary pressures in overheated regions. In moving, companies may well reorganize and become more efficient, possibly taking on additional staff. New companies create spin-off effects for others in the area.

Where investment is brought in from abroad, this may not involve losses elsewhere within the UK unless inward companies compete with existing producers — as is likely to be the case with Sunderland's Nissan car plant, for example. Not all foreign companies will direct investment in the UK to research and development activities. Instead, they may acquire plant dedicated largely to basic assembly. There is evidence that too many British organizations may be seeking out what may prove to be limited overseas investment (Roberts and Noon, 1987).

Although it is easy to be critical of many of the promotional activities undertaken by councils in the industrial field, it is an area many of them consider to be appropriate. It is relatively cheap to implement; it can usefully collate material on, say, the availability of land and premises; and some studies suggest it has been a worthwhile exercise (Storey and Robinson, 1981; Horne, 1988). Whether these types of initiatives make a great deal of difference to what would anyway occur, or whether they are important compared with, for example, the availability or otherwise of regional grant, is open to question.

In attracting in conferences and boosting tourism, different considerations apply. Using under-utilized facilities for conferences may make a great deal of sense, and can promote a city at little cost beyond its immediate region. Purpose-built projects, such as the National Exhibition Centre, have also succeeded in both advertising the benefits of the area and in creating jobs. With other tourist-related activities, however, questions need to be asked about the quality of the jobs being created. Some may be seasonal and poorly paid, but authorities may need to compete in what is an expanding market. Some tourist-related projects may be beyond the financial resources of local government, and will necessitate substantial private-sector investment.

Land and Property

Under legislation such as the Local Authorities (Land) Act 1963, the Town and Country Planning Act 1971 and the Inner Urban Areas Act 1978, local government can intervene in aspects of physical development. They can service land for industrial use and build industrial premises that may be sold, or retained and managed by an authority. They can provide mortgate finance for those wishing to construct industrial units. They can also assist in the improvement of older areas of industrial/commercial property, notably in Industrial Improvement Areas (see Chapter 3).

Intervention in the servicing of land for industrial uses and providing premises for enterprises has been practised for many years by local authorities. Perry and Chalkley (1985) suggest that in the late 1970s and early 1980s, there was a marked increase in the provision of small units by local authorities throughout the country, with the exception of the more prosperous South East and some rural areas. Between 1976 and 1981, well over a million square feet of industrial property was constructed by local councils. In more recent years, cities such as Birmingham (Birmingham Council, 1986) have continued to be very active in the property market by marketing sites for development, by providing a land/ premises package for incoming firms and by improving older areas of industrial buildings. As the major recent analysis of industrial property has made clear (Fothergill, Monk and Perry, 1987), there is every justification for public-sector intervention creating new industrial units because of the outmoded nature of so much existing property. Public-sector intervention may also lower rents, create vacancy chains and enhance efficiency.

However, it is becoming apparent at the time of writing that other public-sector organizations, notably English Estates, are beginning to become far more active in this area. With diminishing local-authority resources it may well be that other agencies will take over a great deal of new industrial building. In some parts of the country, the private sector will continue to develop industrial property. In the more depressed regions, however, it was only during the early 1980s (when advantageous tax regimes applied) that the private sector proved particularly active in the field.

Although local councils may regret the diminishing role they are likely to play in terms of industrial development, there may be some advantages in this. They can concentrate on other under-developed areas, some of which are explored below. It is clear that there has always been tension within many local councils as to the function of local government in the industrial-property market. On the one hand, estates departments typically want to select the safest tenants, thus guaranteeing return on capital investment. On the other hand, other interests represented within local government (usually within planning or economic development departments) tend to adopt different attitudes. They may want to restrict tenants to those producing 'socially-useful' goods, and may want to encourage new, possibly socially-owned, enterprises that may have no proven record. These two positions may be difficult to reconcile.

Business Advice

Sellgren's study (1987) indicates that providing business advice is probably the most significant of all the economic development activities undertaken by local authorities, and is one of the fastest growing. This broad area may encompass many separate initiatives. Authorities may provide, often in co-operation with other organizations, assistance in relation to financial, managerial, technical and marketing issues. They may provide courses in various aspects of business management, usually through local institutions of education. Often, councils will make available information relating to the entire range of financial schemes provided by central and local government and other bodies designed to assist the creation and/or expansion of firms (see, for example, Hammersmith and Fulham London Boroughs, 1987). Providing business-advice facilities may be relatively cheap and easy to introduce. It can help in co-ordinating public and private sectors, as has happened in local enterprise agencies (see Chapter 8). However, supplying pertinent, professional advice requires specific kinds of expertise that may not always be readily available within local government.

Financial Assistance

Some councils will provide financial assistance to companies, or highlight other potential sources of aid. Coventry City Council identifies eleven sources of financial aid available for industrial concerns (1987). Some of these, such as regional assistance, European funds and various urban grants, may not be under the direct control of the authority. Others, including schemes designed to cover training costs, start-ups, business development loans, and so on, are. This type of activity is perhaps, surprisingly, widespread. Mills and Young (1986) found in their survey of over 240 local authorities that half had provided some sort of financial assistance to companies after 1980, and 40 per cent or so had given grant aid. A few authorities, perhaps 15 per cent, were prepared to guarantee loans. Grants and loans had been used for a range of functions, of which the most important was rent/rate relief and the conversion and improvement of land and buildings for industrial use. Total resources allocated to direct financial assistance by the 150 or so councils prepared to give details was insignificant, amounting in 1983–4 prices to little over £22 million.

One debate surrounding the question of financial assistance to companies is what form aid should take. Loans may increase corporate gearing ratios and will obviously have to be repaid, although funds can be recycled. Grants may do nothing to improve the efficiency of firms, and may be simply assimilated by the company to little net benefit. The local authority concerned loses these resources, which may in effect benefit one local producer to the non-benefit of others. For these kinds of reasons an increasing, but still very small number, of authorities have turned to equity funding – that is, investing in the risk side of firms. Money may be lost, but it may increase substantially depending on the fortunes of the company involved. It can also enhance an authority's control over, or influence within, local industry.

Although authorities probably do have powers to invest directly into the equity side of companies, the most interesting developments in this field have come when councils have established separate enterprise boards. The history and development of enterprise boards represent one of the most fascinating aspects of local economic development. Unfortunately, there is not the space here to discuss this at length: those interested should refer to other sources (Centre for Local Economic Strategies, 1986a; Centre for Local Economic Strategies, 1987b; Clarke and Cochrane, 1987; Massey, 1987; Lawless, 1988).

Enterprise boards' (EBs') origins lie in political and economic developments that occurred in the early 1980s. The Labour government had been defeated in the 1979 election, and its efforts to intervene in aspects of national and regional economic development through the National Enterprise Board had proved insignificant (Lawless, 1981b). To an increasing number of radical councillors and officers in some of the major urban authorities, the economic policies implemented by the Conservative government elected in 1979 had to be counteracted locally. National policies designed to reduce inflation imposed severe constraints on industry as a result of high interest rates and a soaring pound. Job losses in manufacturing that in many cities amounted to more than 1,000 a month, necessitated (in some councils' views) direct political action that, in the economic sphere, was designed to increase local government's control over productive capacity. This aim was to be achieved by equity investment effected by EBs.

EBs were separate companies established by local authorities and funded initially through Section 137 of the Local Government Act 1972, which allowed a 2p rate-call to be used for purposes generally beneficial to local residents. Using these resources, five EBs were established in the early 1980s, one – Lancashire Enterprises – by a county council, and four by metropolitan counties: the Greater London Enterprise Board (GLEB); West Yorkshire Enterprise Board (WYEB); Merseyside Enterprise Board (MEB) and the West Midlands Enterprise Board (WMEB). Although similar development companies have been established by other authorities, and the five original EBs have established spin-offs with metropolitan districts after the abolition of the metropolitan counties in 1986, much of the debate and literature still surrounds the five original EBs.

In the early years, there were differences in the strategies adopted by the EBs. GLEB was the most interventionist, investing in its first thirty months over £50 million in equity funds and property acquisition and development. It advocated a number of broader, socio-political goals designed to improve the quality of the jobs it supported, and to enhance the opportunities for the economically marginalized. However, its activities were substantially curtailed in the mid-1980s with the abolition of the Greater London Council, and by having its annual income reduced from £20 million to about £1 million. From about 1986, its policies and programmes were substantially the same as those adopted by the other EBs.

From the mid-1980s on, a number of activities tended to characterize the EBs. The major objective centred on providing equity finance for locally-owned companies, particularly in manufacturing. Assistance has often been subject to a planning agreement covering such issues as trade-union rights, employment levels, health and safety matters, and so on. Some EBs have concentrated on particular sectors of the local economy. Regional development trusts have been established that, as in the case of the WMEB, have attracted substantial private-sector support. Funds have been set aside for small firms, for training initiatives and for aid to technologically-advanced projects.

By 1986, total investments undertaken by the EBs amounted to approximately £50 million. This had been invested in about 350 companies employing almost 15,000 people. Investment per job was low – at between £2,000 and £3,000 per job – although, as it is difficult to estimate total costs or whether jobs would in any case have survived without assistance, such figures need to be treated carefully. Nevertheless, the figure is low when compared with, for example, regional development grant, which creates jobs at costs that may amount to ten times this figure. The EBs have also had very few failures; they have proved profitable; they may well have improved the management of many of the companies assisted; they have negotiated favourable loan arrangements with private banks; they have in part countered the heavy concentration of venture capital in the South East, where almost two-thirds of UK investment was occurring in the mid-1980s; and some, notably the renamed Yorkshire Enterprises, have significantly widened their activities through agreements with local councils.

On the other hand, it is apparent that, increasingly, commercial rather than social objectives have come to dominate EBs. This is perhaps inevitable – there are always going to be conflicts between profits, efficiency and economic survival, on the one side, and broader social ends, such as improving job opportunities for black people and women, on the other. One obvious constraint is that social objectives entail costs. Who will pay for these? But if the kinds of political ends originally assumed by, say, GLEB, have virtually disappeared, the Centre for Local Economic Strategies comment that they have 'pioneered a new type of relationship between local industry and local authorities and between the public and private sectors as a whole, combining day-to-day flexibility and commercial confidentiality with public accountability on broad policy' (1987b, p. 18) seems a fair assessment.

Innovation and Development

Authorities have increased their activities designed to enhance the formation and survival of new firms. Throughout the 1980s, there has been a substantial increase in the numbers of managed workshops – sheltered environments in which a range of technical and professional services are provided for new companies for a short period of time. A 1987 review of managed workspaces (DoE, 1987b) concluded that, although generalizations were difficult because of the wide range of initiatives, this type of project had a clear role to play in helping to create and to sustain small firms. Though not all managed workspaces are under the direct aegis of local authorities, most are, and evidence suggests that policies towards tenure, management, services provided, and so on, can be critical in ensuring the success of such enterprises.

Authorities are looking increasingly at enhancing not only low- or no-tech enterprises (which are typical of many managed workshops) but also at high-tech initiatives. Many northern urban councils, however, find their efforts to develop new industries of only marginal value because of the market's overwhelming attraction to parts of southern England, especially the zone within fifty miles or so of the M4. The reasons for this are addressed elsewhere (Hall *et al.*, 1987), but they include access to Heathrow and the motorway; proximity to government and other research establishments; environmentally-attractive localities; high-quality housing; and so on. Few urban authorities in the older industrialized regions are able to compete with these locational advantages.

Nevertheless, local authorities have created a range of policies designed to boost either new enterprises or – and perhaps of greater significance for many cities – the adoption of new techniques within existing firms. Councils can, for instance, train their own staff in new techniques. They can undertake sectoral studies designed to highlight the need for and availability of new techniques. They can sponsor ITECs designed to provide computer consultancy and training, often with European Social Fund assistance. Product-development centres can be created, such as SCEPTRE in Sheffield, that help to design and to produce new goods and services and that tend to be funded by levies on sales. Microsystem centres have been created to provide impartial advice for local manufacturers. Units have been formed to promote computer-aided design and manufacture (CADCAM). Innovation and development centres have also been established to provide expert advice and access to venture capital for newly-created, in-house firms.

Although these developments are considered important by many local councils, science parks are usually considered to be the ultimate goal. In 1987 there were 36 parks and their development and success have been subject to increasing research (Monck, 1986; Sunman, 1987; Monck *et al.*, 1988). Science parks have formal links with a local institute of higher education, and are designed to encourage the formation and growth of knowledge-based industries, in part through an active management engaged in transferring technological and business skills to tenants. By 1985, about £150 million had been spent on the parks, of which £35 million came from local government. This funding was mainly of property development that, by 1986, housed over 400 firms – many engaged in aspects of computing.

Local government's role in developing science parks has varied considerably. Some have facilitated and promoted them. Others, such as Manchester, have contributed substantially to the physical development of the park, and have been members of the relevant company. Yet others, such as Leicester County Council, have planned, financed and managed schemes (Monck *et al.*, 1988). It can be assumed that many other councils will attempt to facilitate the creation of science parks and/or similar developments: they help academics to create new companies; companies within science parks tend not to displace a great deal of existing output; they assist in diffusing technology; and they exert strong local multiplier effects in sectors such as banking and advertising and, as employers and employees tend to be highly paid, also in leisure, catering and recreation.

As a final point, it is apparent that even where councils have been interventionist, the net effect in many instances of innovative development will be limited. 'High-tech' activities are dominated by market considerations. For many local administrations, Jowitt's suggestion (1988, p. 137) that 'economic development should be premised on the idea of a mixed economy, based on a slimmed-down and technologically advanced assisted manufacturing base, the continued development of the service sector, expanding tourist and leisure services, and lastly a small dash of high-technology' seems appropriate.

Training

The 1980s have seen a considerable increase in training and related activities undertaken by local authorities, often in conjunction with educational establishments, employers and the MSC (renamed the Training Agency in 1988). There are a number of justifications for this development. A lack of skills correlates strongly with unemployment and so enhancing personal skills may improve the economic status of trainees. Unless there are sufficient local jobs, however, those receiving training may simply displace others in the labour market that leads to a reshuffling, rather than a reduction, in unemployment. On the other hand, there may be local skill shortages that restrict output in specific sectors, in which case training can play an important role in removing bottlenecks to production. Additionally, those with skills may ultimately be more likely to establish their own firms, thus potentially creating additional jobs in the longer term.

The range of activities undertaken by local councils that might be seen in part to relate to training is formidable. Unemployment Task Forces have been created that, among other things, offer careers guidance and retraining for those without work (Cheshire County Council, 1987). New-technology training schemes, and short courses appropriate to the technical and managerial needs of locally-important sectors, have been supported (Birmingham City Council, 1987c). A local authority may try to create apprenticeships in the construction industry, through its own direct works section. Information and advice on training provision can be concentrated more directly on those in need. Management and technical training for those about to create their own small firm has been instituted (Coventry City Council, 1987). Some councils have made attempts to use training as a specific mechanism through which to achieve broader social objectives, such as widening job opportunities for women and those from ethnic communities (Islington Council, 1987). This may be done by targeting advice and counselling services to particular groups, by putting on courses directed at certain sectors, by employing trainers with experience of the needs of the disadvantaged, and so on.

Many schemes – and a great deal of local training provision in general – are supported in part by the European Social Fund (ESF), which is particularly biased towards vocational training and job-creation projects for the under twenty-fives. The ESF will fund up to 50 per cent of the expenditure of approved projects, and it can be used as a direct subsidy to encourage employers to take on additional employees. Taking on additional employees has tended to be aimed particularly at the under twenty-fives, and throughout the 1980s has been used by certain councils (Botham, 1984). Some councils, such as Strathclyde, have seen these job subsidies as a mechanism through which a few thousand jobs might be created or sustained. However, as might be anticipated, ESF applications from British local authorities greatly exceed allocations, which tend to be biased towards retraining projects in the depressed regions (Teague, 1987).

Municipal Activities

Local authorities themselves can undertake a series of initiatives designed to improve economic development and/or employment generation. Local government employs almost three million people and, as is discussed below, has a considerable capacity to employ more – given appropriate financial and political incentives. Local government can play a role in training its workforce, in providing apprenticeships in areas such as office work and the construction industry, and in widening job opportunities in its recruitment policies for disadvantaged sectors, such as the disabled.

Many urban councils have established separate departments charged with the task of economic development. As other observers point out, these can vary considerably in terms of staffing levels, in-house expertise, resources, relationship to committee structures, and so on (Mills and Young, 1986). Typically, these departments are involved in activities such as industrial-estate management, product development, equal opportunities and in implementing or helping to initiate the types of schemes mentioned elsewhere in this chapter.

In addition, economic development departments tend to be active in areas such as research, advocacy and policy development. Research can be an important function. Authorities and others need to know about such issues as sectoral change in the economy, skill shortages, the availability of property, and so on. Some authorities have lobbied long and hard for specific changes – for example, Birmingham, after years of endeavour, finally obtained assisted-area status in terms of regional aid. Economic development departments can also produce policy documents relating to areas such as tourism or leisure development. These are important in outlining major trends, and in highlighting the authority's and other agencies' roles in policy development.

Although authorities have traditionally played an important role in employment creation in the cities (where they still tend to be the largest single employer) events towards the end of the 1980s suggested that this was to change. The Local Goverment Act 1988 stipulates that a number of functions normally undertaken by councils, such as refuse collection and street cleaning, have to be placed out for competitive tender. Local authorities can compete for contracts, but they must achieve appropriate financial targets. This Act will certainly lead to a reduction in local-government staff through privatization – a process that will possibly entail net job loss.

However, authorities can still attempt to boost local employment by seeking out local suppliers for goods and services. Many authorities have taken an active role in trying to enhance local job opportunities through contract compliance policies. This is an interesting area that is likely to receive greater attention in the future. Experience indicates, for example, that many construction companies are prepared – indeed, some are eager – to employ people living in deprived areas and to purchase local goods when undertaking specific projects. As those employed may well move out of the inner city during the construction or later operation of retail, commercial and industrial schemes, some cross-over of benefits may occur (Haughton, Peck and Steward, 1987). Nevertheless, as the authorities find their own capacity to initiate schemes diminishing, it will be important for them to try to instil social considerations into projects over which they may have some say in awarding and monitoring contracts, but that are implemented by the market.

Other Activities

Finally, there are a number of policy initiatives councils may adopt that, while not directly aimed at economic development or employment generation, are nevertheless partly designed to achieve these ends. For example, environmental improvements in older urban areas that have an intrinsic aesthetic function may also make the city more attractive to investors. Similarly, developments in transport often have economic overtones: innovations in public transport can ease labour mobility; new or improved road communications are seen as important by employers and investors; and major infrastructural projects such as, for example, new or enlarged airports or the construction of freight terminals for the Channel Tunnel, have employment implications both during and after construction.

Housing policies should also be mentioned. Improving and building dwellings is a particularly effective mechanism through which jobs can be created, particularly when a high proportion of British goods and services are used. Some authorities, notably Liverpool in the early 1980s, placed a considerable emphasis on council intervention in housing, and largely avoided economic development policies in themselves. However, central-government financial controls on housing investment made this strategy less viable in the mid-to-late 1980s. One housing policy that many authorities continue to implement is that of key-worker housing: employees considered as highly important to incoming or expanding firms may, under this scheme, be offered housing on a temporary or permanent basis by the local authority concerned (see, for example, Metropolitan Borough of Knowsley, undated).

Local Economic Development: Debates and Tensions

Earlier sections of this chapter have only hinted at the types of initiatives local government has introduced, largely since 1979. It is important that observers of British urban regeneration are aware of the variety and richness of locally-effected policies, not least because central governments elected since 1979 have undoubtedly played down the role and impact of local government in urban regeneration. Nevertheless, this range of economic development policies must not detract from broader considerations. Two major themes merit study here: central–local government tensions inherent in local economic development; and the dilemmas of local economic development.

Central–Local Relations in Economic Development

Since 1979, relationships between central and local government have been one of the most dynamic influences in the ordering of inner-urban intervention. Strong arguments suggest that one of central government's major motivations in effecting an inner-urban policy has been to assimilate powers previously within the sphere of local authorities. These arguments surround debates about the origins of and explanations for inner-urban policy, which is examined in greater detail in Chapter 10. However, some mention ought to be made here of the specific nature of central–local relations as they apply to local economic development, since these are complex and – from central government's point of view – are ultimately intended to diminish the scale and scope of local intervention.

Central governments elected since 1979 have generally been antagonistic towards local economic development. Public-sector intervention at a local level is not attractive to Conservative governments committed to a substantial curtailment and not enhancement of State activity. Official statements, such as *Lifting the Burden* (HMSO, 1985), reveal a government determined to reduce regulation and control undertaken by both central and local government. It is not surprising that local economic development – an area not traditionally at the centre of local government activity and one imposing few obligations on local councils – should have been the subject of particular attention on the part of recent Conservative governments. It may well be argued that no government since 1945 has been especially supportive of local economic development. One reason for this could be that, even if at the margins, local economic development can be implemented according to different economic objectives from those pursued by national administrations. No Chancellor has been eager to decentralize economic powers. Even with changing political climates, there can be no guarantee that one ever would.

Although in the 1980s central governments have not been in favour of some of the local initiatives outlined earlier in this chapter, the Department of the Environment (in advice contained within circulars) has indicated that local government has a specific and limited role to play in industrial development and employment creation (DoE, 1984; 1985a). For example, local government should, according to central-government advice, order its planning functions so that any applications involving enhanced economic development are dealt with speedily. Equally, the Department of the Environment envisages councils allocating and servicing land for industrial use. Few could argue with this advice, but it is largely irrelevant to the broader debate: to what extent will central government impose direct restrictions on councils in terms of local economic development and employment generation?

Local economic development rests on a number of legal and financial measures, the majority of which are mentioned above in relation to specific strands of local intervention. Local government has, in general, been innovative in its use of a wide range of permissive powers and financial resources to undertake economic development programmes. Acts such as the Local Authorities (Land) Act 1963, the Inner Urban Areas Act 1978 and the Local Government Act 1972 have been used to grant local councils the powers and resources to operate appropriate programmes. Financially, resources emanating from the MSC, the European Regional Development and Social Funds, historic building grants, tourism grants, Derelict Land Grant and funds from central government's inner-urban policy have all been utilized (Parker, 1985). Increasingly, enterprise boards have been able to negotiate funds from the market. Nevertheless, despite this apparent plethora of potential financial resources, about 40 per cent of spending on English local economic development has come from Section 137 of the Local Government Act 1972. The evolution of this power in the 1980s epitomizes particularly well the tensions between, on the one hand, central government's determination to limit local intervention and, on the other hand, local government's concern to heighten economic development.

In 1980, the Conservative government established a review body on local economic development under the chairmanship of the Department of the Environment's Chief Planner, Sir Wilfred Burns. Although local government had been hoping for a widening and clarification of local economic development legislation, the Burns Committee – in its consideration of Section 137 – argued for a substantial curtailment of powers. Section 137 allowed local authorities to spend up to 2p of the rates on expenditure that would benefit the area or some or all of all of the residents. By the mid-1980s, about two-thirds of all spending under Section 137 was directed towards local economic development. However, the recommendations of the Burns Committee (which were largely supported by the government) were amended in the Lords and, in the resulting Local Government (Miscellaneous Provisions) Act 1983, Section 137's ability to fund economic development proposals was confirmed. But this was not the end of the story.

Although Section 137 was confirmed in 1983, a number of further developments weakened its ability to sustain local economic development. The abolition of the metropolitan counties effectively halved the resources available within the major cities, since one tier of local administration was lost. Moreover, in 1985 the Leicester decision ensured that initiatives such as Employment Departments created through Section 137 expenditure would have to be fully funded, including staff costs, out of the 2p provision (Crawford, 1986).

Finally, in 1986, the Widdicombe Committee of Inquiry's report, *The Conduct of Local Authority Business*, was published (HMSO, 1986b). As part of its brief, the committee explored economic development activities implemented under the Local Goverment Act 1972. In practice, the committee was somewhat ambivalent, but, if anything, it was perhaps supportive of local economic development. In the government's response to the Widdicombe Report, however, it was again made clear that central government had grave doubts about the role of local government in economic development (HMSO, 1988b). It argued that some activities would be better left to the private sector or other public bodies; that local economic development might simply involve a migration of economic activity from one place to another; that some land acquisition projects appeared overly expensive; that cost-effectiveness was by no means assured; that projects may well have gone ahead without local government support, and so on. Nevertheless, as some projects were regarded by central government as having intrinsic merit, it appeared in 1988 that a general, circumscribed power will be introduced allowing local government some powers in the field, but removing economic development from other legislation – including Section 137. Certain activities, such as commercial trading and equity investment, will not be allowed. However, local authorities will still be able to form separate companies for specific purposes.

This brief summary of the complex development of Section 137 in the 1980s illustrates a number of the tensions inherent in central–local relations in urban regeneration. It highlights the contrasting objectives adopted by the two tiers of government. The centre perceives local economic development as a marginal activity, where councils provide in the main a physical framework for the market. As this chapter has already shown, local government has viewed the area in an altogether more imaginative light. Not surprisingly, this lack of common ground has led to what Young (1986b) has called a vacuum in central–local relations. At the time of writing, it would appear that this vacuum will be filled by central-government inspired legislation that seems likely to inhibit the scale and scope of local innovation.

Central goverment's determination to limit the scale of local-authority activity in economic development and employment generation appears misguided. Local-government resources allocated to economic development are small: they amounted to about £220 million in 1986–7, about 1 per cent of total local-government expenditure (HMSO, 1988b). Local government has shown a degree of innovation and imagination that has not been paralleled by central government. Councils are in a far better position to implement economic strategies within the cities because of their knowledge of, and contacts with, such interested parties as employers, community groups, financial institutions, and so on (Association of Metropolitan Authorities, 1986). It also becomes increasingly apparent that local economic development is particularly cost-effective when compared to nationally-instituted policies.

As the whole question of evaluating instruments of urban regeneration is an important one, and as it impinges on many aspects of inner-urban policy, it is examined in its entirety in Chapter 9. Suffice it to say here that evidence suggests that jobs can be created using local-government inspired initiatives, such as enterprise boards, at much lower costs than have prevailed, say, with enterprise zones and regional development grant (Centre for Local Economic Strategies, 1987a). Sometimes, job-creation costs through local-authority initiatives appear to amount to only a tenth or so of resources required through nationally-implemented policies. There seems to be a real need for local intervention – many companies appear badly managed and financially unstable, in need of considerable financial and technical support emanating from a structured, well-researched programme of public intervention (Turok, 1988).

There is one other reason why central government should look more carefully at local initiatives: they can be very effective in creating a substantial number of jobs. This debate was thrown into focus during the 1987 election campaign, when a number of councils produced employment plans (Manchester City Council, 1987; Sheffield City Council, 1987). These made it plain that many urban authorities were in urgent need of additional staff in such areas as social services, housing, training, leisure, education, and so on. Sheffield calculated that it could create well over 10,000 jobs in two years. Some positive assessments of Local Jobs Plans were published (for example, Campbell *et al.*, 1987). These plans would, if implemented, be a far more effective mechanism for creating jobs than, say, reducing income tax. Jobs could be created relatively more cheaply, especially if targeted at the unemployed, every one of whom cost the Exchequer in the order of £6,000–£7,000 per annum in the mid-1980s. As almost every observer of Britain's inner cities has commented on in the 1980s (see, for example, Town and Country Planning Association, 1986; Confederation of British Industry, 1988), there is a desperate need for social investment in the older urban cores. However, events in 1987 and 1988 showed the dominant political realities: the Conservatives won a large majority, with just 43 per cent of the national vote; local-government resources have not been increased to accommodate additional staff; and the 1988 Budget adopted policies that are unlikely to benefit British employment and output prospects – substantial tax cuts were granted to the better-off instead.

Dilemmas of Local Economic Development

Local economic development raises a number of dilemmas. Some of these can be regarded as operational considerations, while others highlight fundamental political issues. The all-pervading political dilemma facing some councils has been the degree to which local economic development is intended to ease constraints on the market, or is designed to counteract its anti-social tendencies. For most local authorities, however – possibly the majority – the objective of local economic development is not a political issue, but simply one of creating jobs. If that means co-operation with other public organizations and the market, then there will be co-operation. On the other hand, the major drive towards local economic development in the early 1980s was fuelled by radical sentiments, based on different political assumptions.

The factors responsible for an intensification in radical local-economic policies are well documented (Cochrane, 1983; Clarke and Cochrane, 1987). Among the more obvious issues were the Labour Party's failure in the 1970s to restructure industry effectively to the benefit of labour; the economic policies pursued by the 1979 incoming Conservative government, which adversely effected much of manufacturing through high interest rates and a over-valued pound; the radicalization of some Labour authorities in the late 1970s and early 1980s; the over-concentration of economic and political power in London and the South East; and so on.

These types of forces encouraged some urban authorities to innovate in local economic development. This intervention was designed, among other things, to instil social as well as economic considerations into aspects of production by, for example, trying to create jobs for the disadvantaged; stressing the role of the local authority in economic development and employment generation; and presenting a series of innovative and progressive policies that would result in the production of socially-useful products (Cochrane, 1986).

In practice, not many authorities were determined to pursue what might be termed radical programmes. Some of the metropolitan counties, one or two cities (of which Sheffield is the best example) and a few Labour inner-London boroughs were in the vanguard. Other Labour councils, while strongly opposed to the national economic policies adopted by Conservative governments elected in 1979 and 1983, were more pragmatic in their approach towards the market. Liverpool, for example, argued that attempting to intervene in aspects of production would have a minimal effect, and could possibly lose rather than create jobs because of increased efficiency, and which in turn would boost industrial profits. Whatever the merits of this argument, Liverpool's job-creation strategy, designed to increase public investment in consumption (notably through the building of new housing and the improvement of older property) seems more likely to sustain employment than would intervention in production.

However, Liverpool's position was not adopted by other radical authorities. Other authorities began to implement a series of policies intended to enhance public-sector intervention in – and control over – production. Local-government economic development initiatives have been outlined above, but it might generally be argued that a number of programmes in particular epitomized radical intervention (Benington, 1985; 1986). Enterprise boards were much to the fore, as they allowed direct investment in production. Some were based on the assumption that support would only occur with the signing of planning agreements with assisted firms that covered such issues as employment levels, trade-union rights, and so on.

The role of the local authority in employing, training, innovating, researching and purchasing was seen as extremely important. Efforts were made to provide research facilities and resources for trade unions, especially in campaigns designed to prevent job losses. New forms of social ownership, such as co-operatives, were supported. Intervention was frequently regarded as most likely to succeed if it was directed towards specific sectors where, for example, lack of investment could be identified. The *London Industrial Strategy*, published in 1983 by the Greater London Council, is perhaps the best example of this kind of sectoral analysis (Eisenschitz and North, 1986). Finally, innovations in product development that a number of authorities effected should be mentioned. For example, Sheffield, in conjunction with local educational institutions, created SCEPTRE, whose early developments – notably a dehumidifier produced by a local co-operative for use in public housing – can be regarded as a notable example of social innovation in production.

Despite these developments, by the mid-to-late 1980s, the underlying economic and political realities imposed upon, and eventually adopted by, previously-radical authorities led ultimately to an effective abandonment of interventionist strategies. Virtually all the projects devised in the early years of the decade were to survive, although often in a modified form, but the political and economic objectives underlying intervention were to change dramatically.

Many of the local economic initiatives designed in the late 1970s and early 1980s were seen as paradigms or parables to be adopted nationally by the Labour Party (Benington, 1986). By the mid-1980s, however, the Labour Party no longer seemed interested in the more radical messages emanating from local parties. And to many observers, Conservative rule began to look semi-permanent. This had obvious implications for previously-radical authorities. Continued anti-market or even neutral stances appeared to hold out little prospect of economic development when local-government resources for economic development were so small and likely to diminish, as the debate on Section 137 makes clear. This was also at a time when pension and insurance funds were cash-rich, which some cities were benefiting from through commercial and retail development. Opportunities for local-authority economic intervention were reduced in 1986 on the abolition of the metropolitan counties. Finally, it has to be noted that the era of radical economic intervention did not result in an extensive restructuring of local economies to the benefit of labour. The impact of radical policies was marginal, even within the handful of councils committed to extensive intervention, and the resources available to achieve this end were extremely inadequate. Nevertheless, the early 1980s were important in terms of local economic development: many new programmes were devised that eventually both Conservative and Labour authorities have come to adopt, and both central government (HMSO, 1988b) and the market (CBI, 1988) have ultimately accepted that local goverment has a limited but valid role to play in local economic development.

Despite this somewhat begrudging acceptance, it remains clear that a number of operational constraints and dilemmas will continue to impinge on policy. A number of these can be identified, but perhaps four can be mentioned here: economic development versus employment creation; the role of small firms; manufacturing employment and alternatives; and the impact of local economic development.

One of the most difficult problems confronting local authorities eager to embark on policies is the degree to which, on the one hand, they are attempting to create jobs or, on the other hand, to enhance economic development. The two are not mutually incompatible, but it is important to note that many policies adopted by local authorities may marginally enhance the efficiency, profitability and/or output of firms but will have no effect on job creation. Indeed, they can lead to job losses. The most notable example might be an enterprise board's equity invest-ment into a firm that needs substantial restructuring, and hence job loss, to survive. One could argue that, even if this occurs, it is preferable for a company to survive and to remain in local control than be lost altogether. However, if jobs are the primary objective, policies such as enhancing community development (see Chapter 8), investment in social, educational and leisure services, and expenditure on housing and infrastructural projects will have more beneficial impact than will corporate grants, loans and equity investments.

A second problem inherent in local economic development is the degree to which it ought to be primarily concerned with the fortunes of smaller companies. The Burns Committee (1980) in the early 1980s, for example, in its consideration of Section 137 of the Local Government Act 1972, argued that the 2p rate should be used primarily to support smaller firms employing less than twenty-five people. In practice, most policies emanating from local government have tended to benefit smaller rather than larger companies. The logic of this is presumably that smaller firms may need more help. Moreover, if they survive they will help to create and to sustain more local jobs within locally-controlled firms. However, the validity of this has rightly been open to considerable debate (Greater London Council, 1983). Small firms create few net new jobs; they often pay lower wages than larger companies and provide far fewer labour benefits; they may often simply provide the same goods as larger companies; and, importantly, they are often dependent on larger firms for their contracts. No doubt local authorities will continue to be encouraged by central government to concentrate their support on smaller firms, but it is the large- and medium-sized sectors that employ far more people. Many medium-sized, locally-owned firms need technical, financial and training assistance as much as smaller companies. Their fate is ultimately far more important to the urban economies than are smaller concerns. However, this is unlikely to be reflected in many local economic policies in the 1990s.

A third issue of concern is the extent to which local economic development should support manufacturing rather than other forms of employment. Traditionally, many urban authorities have tended to bias their support, if only implicitly, towards manufacturing. There are a number of reasons for this, including the dramatic loss of manufacturing jobs in the early 1980s, the importance of trade unions in industrial production and, within some Labour urban councils, a kind of moral argument that manufacturing jobs are 'real' jobs (Bruegel, 1987). By the end of the 1980s these attitudes had weakened, and it can be assumed that they will continue to do so. Less than one-third of those in employment work in industrial production and, in many cities, the big expansion in terms of physical development has occurred in the retailing and commercial sectors, with some associated employment growth. Many service sectors primarily serve local populations, but some producer services, for example, banking, insurance and computing, can export goods and services and thus boost local employment.

Councils need to look increasingly at these sectors for local job creation. How councils are to do this is a more difficult question. To a large extent, they may need to revert to older methods: promotion, advertising, seeking market investment and providing basic infrastructural investment. More radical policies are unlikely to appeal to the major institutional investors who are responsible for most development. Recent years have shown that financial backing for retail projects can more or less be obtained throughout the country. Commercial (and particularly industrial) investment is much more difficult to obtain outside the South East. Authorities wishing to see these kinds of developments within their administration will find themselves involved in issues of image and market appeal.

To return to the important constraint that undermines local economic development: its marginality. The scale of resources allocated to this area ensure that local government will not play an interventionist role in local production. It may be argued that this is correct, that councils do not have adequate expertise in finance, management, production and marketing to justify any substantial role in production. In practice, however, enterprise boards have shown that many medium-sized companies do need financial and managerial assistance. Nevertheless, for many councils, local economic development programmes will concentrate increasingly on devising and implementing limited programmes of financial, technical and training support for small firms (Teitz, 1987). The link between attempting to promote economic development and trying to widen job opportunities for the disadvantaged (central to the strategies created by more radical councils in the early 1980s) has been lost (Cochrane, 1988). Whether all of this makes sense remains open to considerable debate.

Trying to enhance output and development, which many councils will seek to achieve, will be increasingly difficult when urban economies are so fragmented. Thriving retail sectors can exist at a time when manufacturing continues to decline. Attempting to ensure an across-the-board boost to the local economy, may be impossible for authorities to effect. National and international agencies alone can implement policies likely to increase output in an era when production is itself so internationalized. Perhaps, therefore, local authorities would be better advised to concentrate on labour-market strategies that are designed to ensure a fairer distribution of jobs. (Lovering, 1988).

Local Economic Development: Conclusions

In 1988, the whole question of local economic development was thrown into some confusion by the publication of a Department of the Environment Consultation Paper on local authorities' interests in companies (DoE, 1988c) and the goverment's response to the Widdicombe Report (HMSO, 1988b). Although at the time of writing the full impact of these documents is uncertain, it would appear that substantial changes to the legal basis underlying local economic development may well be introduced in the late 1980s. As has been mentioned, a new, circumscribed power to undertake local economic development is likely to be introduced. In addition, it would appear that companies controlled by local authorities may have problems in undertaking commercial activities and in holding equity, and may well have their capital and current expenditure amalgamated with that of the parent council. These proposals are likely to have a severe impact on the scale and direction of local economic activity undertaken by both councils and by separate public–private-sector developments.

8
Non-Governmental Institutions and Inner-Urban Policy

The previous five chapters attempt to provide a review of most of the major inner-city initiatives formulated by central government or by local authorities. In this final chapter of Part II, the emphasis changes to non-governmental agencies. As the chapter shows, this is something of an eclectic categorization.

Basically, these agencies are essentially independent organizations, although they tend to retain or develop strong links with either local or central government or both. They usually adopt inner-urban regeneration as one (although by no means always the most important) of their objectives. Perhaps most importantly, one characteristic common to many of them is that they are growing in influence. Ten years ago many of the approaches and institutions discussed here either did not exist, or were much less important than they have become towards the end of the 1980s. Classification in this area is far from easy, but three main influences can be identified: trade-union activities, community and co-operative enterprise and private-sector initiatives.

Trade-Union Initiatives

Trade unions have fostered inner-urban regeneration in at least three ways. First, they have initiated or contributed to local economic developments (LEDIS, 1986). For example, they have created specific resource centres and allocated funds to assist in the creation of workers' co-operatives. They have been involved in the management of community development projects, sometimes in collaboration with local businesses. Local development agencies often have trade-union representation alongside corporate and local-government interests. Union officials have also been seconded to youth projects and community ventures.

Second, the Unity Bank (created in 1984) has become an important mechanism through which trade unions have assisted in urban development. The Unity Trust, which controls the bank, is a public limited company and a subsidiary of the Co-operative Bank. It has substantial support from more than fifty trade unions. By 1986, its resources amounted to about £10 million in share capital and £60 million in deposits. Its activities are varied, but include issuing regional invest- ment bonds that are sold to individuals and institutions to provide loan finance for local industrial concerns. It is also attempting to stimulate various forms of employee share-ownership plans, either through providing advice or direct financial support. With the increasing centralization of corporate power in the south of England, employee share-ownership schemes offer important pos- sibilities in that they encourage local workers to take a direct interest in (and even potential control over) the enterprises in which they are employed.

Third, trade unions have helped to stimulate urban regeneration through lobbying, mainly by issuing publications and the publicity this generates. In 1988, for example, the TUC argued for a new approach to the inner cities which was based on a national strategy aimed at creating more highly-paid, higher- productivity service jobs that, through training, would be available for urban residents. Additionally, the TUC argued for increased investment in such areas as transport and education, and for a rejection of the idea that private-sector investment in the cities would ultimately lead to a trickle–down of benefits to the less well-off – an assessment that appears to be generally accurate in the context of British inner-urban policy. Instead of private-sector initiatives in the cities, all interested bodies (including local and central government) should be involved in devising and implementing a comprehensive urban strategy (TUC, 1988).

Co-operatives and Community Enterprise

In the 1980s, there has been a remarkable increase in interest in co-operative and community enterprises. There are a number of reasons why this has occurred. Increasing unemployment has forced many people to explore alternative ap- proaches towards job creation. The tendency for larger employers to reduce their workforces in peripheral regions has encouraged ideas of local initiative and control. More people are interested in pursuing less-orthodox life-styles that may well be more compatible with the rhythms of community business than with the rigours of large-scale enterprises. The 'small is beautiful' syndrome has also helped. However, whatever the motives, the overall impact has not been very great.

Not all community-based initiatives and co-operatives have been located in cities. However, many have, and any analysis of urban policy would need to include these kinds of initiatives. One way into this complex area is to consider co-operatives and community enterprises in the following manner: definitions and objectives; evolution and development; and evaluation.

Definitions and Objectives

Co-operatives provide services and/or manufacture goods, and they are owned and controlled by those working in them. They are autonomous enterprises in which employees become members by holding share capital. Members share in profits and usually have some say in decision-making in the organization, normally on the basis of one person, one vote. Returns on capital are limited. Variations in structure and organization do occur, although a large number have adopted model rules laid down by the Industrial Common Ownership Movement (ICOM), the national umbrella organization.

Co-operatives' aims inevitably differ. However, they generally tend to include both economic and social aims (Thomas, 1988). Economically, goals such as job creation and/or retention may be important, but other socio-political ends, such as the desire to work within a democratically-controlled enterprise, may be just as significant to the members.

Community enterprise is more difficult to define. One problem is the use of the wide-meaning word 'community', which can mean at least three different things in the context of urban renewal. It may, for example, be used to describe activities undertaken by the voluntary sector, which are examined elsewhere (Marsh, 1988). These projects include youth-training schemes, technology training-centres, women's groups, and so on. They tend to be funded by the Urban Programme (see Chapters 3 and 4). Two other community-based approaches deserve greater consideration: development trusts and community business.

Development trusts are independent, non-profit-making organizations that attempt to integrate public, private and voluntary sectors into projects intended to regenerate economically and socially specific places or schemes. They often bring together financial and other resources from a wide range of organizations. Many of them, for example, acquire resources from local government, the Urban Programme and the private sector. Their inclusion here is not inappropriate, since they try to incorporate local individuals and organizations to achieve identifiable community gains.

Development trusts were subject to a major review in 1988 (DoE, 1988a). Among the more pertinent findings are that development trusts can often achieve results in difficult circumstances by carrying out improvements that meet local objectives, often using local resources. They can help to reinforce social links in – and act as focuses for – local areas. Good-practice principles seem to apply, including an early identification of clear objectives, the appointment of high-calibre executives, the incorporation of management skills in a trust, an ability to synthesise support from different sources and the effective use of local communities. It may be far from easy to sustain all of these desirable facets, but those familiar with the work of the more successful development trusts will appreciate how dramatically they help to instil social and physical change in local communities. The work of the Wirksworth Project in Derbyshire, for example, in renovating a decaying town, or the Eldonian Community Trust in constructing co-operative housing in Liverpool or the Pennine Heritage in helping in the reconstruction of Hebden Bridge (W. Yorks.) is very illuminating. The potential capacity of development trusts has not been exploited to the full.

Although innovations such as development trusts can be regarded as aspects of community enterprise, much of the debate in this area has concentrated on what are usually called community businesses. As with development trusts, definitions can be a problem, but some characteristics seem to be held in common (Buchanan, 1986). They are trading organizations that are community based and owned. Residents are entitled to become members upon the payment of a small annual subscription. Trading profits are used either for re-investment in the enterprise, as a bonus to workers, for some other community benefit or, if a subsidiary of a charitable company, paid to the holding company for use elsewhere. Outside expertise, in such a form as local councillors and those with management experience, can also be co-opted onto boards. Community business activities are wide ranging, and include workspace provision, construction companies, craft production and food co-operatives. Some are involved in running a particular business, while others are concerned with trying to stimulate and support other community enterprises that are still in their early stages. Ultimately, however, community business 'sets out to be a model of, and provide a structure for, community power, community control, community ownership, and community autonomy' (Buchanan, 1986, p. 19).

Co-operatives and Community Enterprise: Evolution and Development

The recent history of the co-operative movement has been one of considerable expansion. In the 1970s, the Labour government assisted in this process by, among other things, passing the Industrial Common Ownership Act 1976 that provides a legal definition and a relatively-small loan fund; the government helping in 1978 in the creation of the Co-operative Development Agency, which is designed to enhance the formation and success of co-ops; and formulating, in 1978, Inner Urban Areas that allow urban authorities to make small grants and/ or loans to those wishing to establish co-operatives (Whyatt, 1983).

This kind of infrastructural support, combined with far greater assistance from local government and a somewhat reluctant all-party acceptance of the principle of co-operative ownership, has helped the movement to expand considerably, although it is difficult to establish exactly to what figure. By 1987, some estimates indicated that there were about 700 registered co-ops in Britain – a sevenfold increase on the 1978 figure (Centre for Local Economic Strategies, 1986b; 1987c). Other observers put the figure much higher than this, one suggesting that 1,300 co-ops were in existence by 1987, with a total membership of about 20,000 (Batchelor, 1987).

Based on the Centre for Local Economic Strategies' lower figure (1986b; 1987c), certain key features characterize the British co-operative sector towards the end of the 1980s. Although the 670 or so co-ops employed about 5,100 full-time and 1,560 part-time workers, these figures are distorted by five large manufacturing co-operatives that collectively employ more than 1,000 full-time people. If these are excluded, the average size of each co-op is about 3.3 full-time and 2.2 part-time people. Almost a half of all co-ops are in the service sector, although more full-time employees are in manufacturing – again because of the influence of the small number of large industrial co-ops. Almost one-third of co-ops are in London, with low numbers in such regions as East Anglia and the South West.

Commentators have identified a range of social, economic and political processes that have, in some way or other, influenced the development of the movement (Centre for Local Economic Strategies, 1987c). Among the more important processes are those when an owner donates or sells a going business to workers who run the enterprise as a co-op; when workers from a business that has either closed or contracted set themselves up as a co-op, sometimes in competition with the parent firm; or when individuals (sometimes from marginalized backgrounds) come together to create a co-op in a sector that usually requires little initial financial capital. Whatever the processes, the net result has been a considerable expansion in co-ops from a very small base in the early 1970s, to what, by the late 1980s, was still an insignificant element in total national employment.

Unlike co-operatives, which have been developed throughout the country, community businesses have been particularly prevalent in Scotland, where about 100 such enterprises were created in the 1980s, almost all in deprived locations, such as local-authority estates or inner-city areas. One reason for the scale of the Scottish activities is the degree of support from a variety of national, regional and local organizations (Hayton, 1985). The most influential of these is Strathclyde Community Business Ltd, an independent one-door agency designed to develop and to finance community businesses in the region. It has attracted widespread support from a range of local- and central-government organizations and the private sector (Buchanan, 1986). It also provides a number of different grants to cover such functions as start-up and venture capital, from a budget that, in 1986–7, amounted to over £600,000.

It has been argued that there are several distinct types of community business (McArthur, 1986). Some have been concerned primarily with providing temporary employment and training, usually using MSC funding. Others are commercial trading enterprises, for example, those acquiring and refurbishing furniture and other household goods for subsequent re-sale. Some community businesses are home based, mainly concentrating on knitted goods. Others are development agencies providing accommodation for rented units. However, in general they are designed to create and to retain local wealth through the implementation of schemes that aim to be both economically efficient and socially responsible. Community businesses 'are laying the ground for a new approach to economic development based on stimulating economic activity around accountable community structures committed to the welfare of workers and local people' (McArthur, 1986, p. 98).

Co-operatives and Community Enterprises: an Evaluation

Questions of evaluation in relation to community and co-operative enterprises can be divided into three major areas: economic criteria, social issues and support mechanisms.

Economic criteria

As has already been mentioned, co-operatives and community businesses do not, overall, employ many people. However, by some economic criteria these kinds of enterprises perform well in relation to more orthodox establishments (Buchanan, 1986; Thomas, 1988). For example, failure rates in co-ops were lower than for businesses as a whole until the mid-1980s when, with so many new co-ops being formed, failure rates rose sharply to levels in excess of those within the whole economy. A similar, largely-positive evaluation of community enterprises in Strathclyde pointed out that in 1986, 35 trading community businesses had directly or indirectly created 1,800 jobs, although of these only 10 per cent were full-time workers employed directly by a community business (Buchanan, 1986). The remainder were part-time jobs, training places, and so on.

By other economic indicators, co-ops and community businesses do not fare as well. The failure of the so-called 'Benn' co-operatives in the 1970s – KME, the *Scottish Daily News* and Triumph Meriden – highlighted the problems many co-ops and allied ventures encounter in the commercial market. For example, they tend to be chronically undercapitalized (Mellor, Stirling and Hannah, 1986). They often operate in low value-added, labour-intensive sectors, where entry is easy and competition fierce. They tend to lack managerial and technical expertise, particularly in areas such as marketing and finance. This can create complications when one or more monopoly customers impose lower prices and tighter deadlines than would normally be the case. The result of these processes is that wages and productivity tend to be lower – often much lower – than is the case with orthodox firms (Buchanan, 1986).

Social issues

By their very nature, co-operatives and community enterprises cannot be evaluated in narrow economic terms. Their objectives are wider than those adopted by most companies. Social rather than economic auditing may be the more appropriate system by which to assess co-ops and similar projects (McArthur, 1986). For example, positive achievements could include an instilling of community pride; a reduction in the costs of public services in such areas as policing, social work and health care; an enhancement of other forms of investment in an area; a provision of economic opportunities for those who would not otherwise find jobs; and help for individuals to gain confidence in themselves by working in a more enriching and supportive environment. These considerations can be important, but they are rarely incorporated into programme evaluation.

However, the kinds of social and community objectives assumed by many co-ops and community businesses can be difficult to achieve, and may contain intrinsic dilemmas for those working in, or supportive of, social enterprises. For example, co-ops may want to be owned and controlled by workers, but this can often lead to commercial difficulties in that in order to compete in the market, additional financial support may be needed by way of loans, grants or equity investment. The acceptance of some of these will undermine autonomy. Equally, although there are opportunities for personal development, it is clear that the pressures of initiating and running an enterprise can be tremendously time-consuming and can impose severe stress.

Support mechanisms

A good deal of the debate about the role of co-ops and community enterprises within urban regeneration has been concerned with support mechanisms (Moore and Skinner, 1984; NCVO *et al.* Working Party, 1984; Hayton, 1985; Centre for Local Economic Strategy, 1987c; Thomas, 1988). From these findings, a general consensus of opinion can be observed surrounding a number of issues. Many community-based initiatives require support across a range of technical, financial and managerial functions. Support needs to be sensitively attuned to the needs and aspirations of social enterprises, and ought usually to be continuous. Support (most of which comes from local government) should ideally be inter-departmental and be based on the identification of key, trained staff. It might also include the formation of a specific community enterprise budget, and the secondment of appropriate staff to community businesses or support agencies.

In practice, a degree of support is forthcoming. In Scotland, as has already been mentioned, community business has received encouragement and support from central and local government. Throughout the country, co-operatives have also received considerable assistance from various support organizations. By 1984 there were more than eighty of these, most of which were either independent, local-authority funded or voluntary co-operative development agencies (Cornforth and Lewis, 1985). Support mechanisms tended to be strongest in areas that had high levels of unemployment, or were suffering from multiple deprivation. In the early 1980s, the Greater London Enterprise Board (see Chapter 7) was particularly influential in enhancing the formation of co-ops in the capital in that it invested in more than 100 co-ops between 1983 and 1986.

There seems little doubt that, where support organizations are installed, co-operative-formation rates increase substantially (Cornforth and Lewis, 1985). Between 1980 and 1982, for instance, areas with co-operative development agencies appeared to create co-ops at five times the rate apparent within areas without support. Jobs within co-ops increased much more in supported areas, and failure rates tended to be lower. Moreover, the costs of job creation appeared low. Each development worker tended to help create about two co-ops, each employing on average four people each year (Thomas, 1988). This may not appear very significant, but overheads within support agencies are low and about two-thirds of the jobs survive for more than three years. Hence, by the mid-1980s, figures per job created were amounting to a very impressive figure of £3,000 to £6,000. This is, in effect, less than it would cost to keep one person unemployed. Not surprisingly, many Labour authorities in particular have supported co-operative development agencies. The agencies have created or sustained jobs very cost-effectively within democratically-controlled enterprises generally committed to broader socio-political goals, such as equal opportunities and the enhancement of personal skills.

Nevertheless, co-operative development agencies and other supporting organizations may run into problems with the clients they are designed to assist. Objectives may differ – local authorities may generally be concerned with job creation, co-operatives with such issues as the quality of the working environment. This diversity in objective definition can lead to conflict where a development agency is too overbearing in its attitudes. There may be too many supportive bodies in some contexts, notably in the case of community business in Scotland. Some bodies, such as banks, will want different things from a community business than will, for example, a local authority. Trade unions may have poor views of co-operatives where, for instance, socially-owned enterprises do not keep to local wage levels – an important consideration for Labour-controlled councils.

More fundamental issues are at stake. For example, to what extent will public subsidy for a co-op or community enterprise displace activity elsewhere? This may be the case with some social enterprises, since their low-tech nature suggests that they often compete with other local firms. Finally, there is the question of the effectiveness of subsidy (Jacobs, 1986). If subsidy is directed towards enterprises that have adopted social as well as economic objectives, it may have to be accepted that similar resources granted to orthodox firms would actually create more jobs. All kinds of issues are raised by this dilemma, including the important question of jobs for whom. Like all subsidies, it is important to define at the outset the scale and aim of support.

Co-operative and Community Enterprise: Conclusions

Increasing co-operative and community enterprise should be a major aim in inner-urban policy. Some encouraging developments have occurred in the 1980s, but the scale of co-operative enterprise achieved in other parts of Europe, notably in Italy and Basque Spain, make recent increases in UK activity appear small by comparison. Co-operatives and community businesses raise problems, but these should not discourage innovation and initiatives in the area – the benefits are potentially enormous. Social enterprises in particular provide new approaches to ownership of production; they allow the disadvantaged some hopes of material advance; they create jobs cost-effectively; and they ensure the local control of enterprises.

Private-Sector Initiatives and Urban Regeneration

Towards the end of the 1980s, the marked characteristic of Conservative inner-urban policy was the determination on the part of government to incorporate private-sector interests into urban regeneration. The 1988 *Action for Cities* statement argued, for example, that 'the inner cities must be places where businessmen want to invest' (HMSO, 1988a, p. 3), an objective that was to be achieved through financial and other incentives designed to attract private development into the cities.

The assimilation of the private sector into inner-city policy reflects a noticeable change in attitudes. In the early 1980s, most commentators saw little evidence of private-sector interest in the renewal of the older urban cores (Lawless, 1981a; Home, 1982). As far as could then be ascertained, industrial, retail, residential and, increasingly, commercial markets were decentralizing from the cities and it appeared as if very substantial subsidy would be needed before private invest-ments would willingly re-invest on any significant scale within the major conurbations outside London.

However, in the late 1980s, attitudes and policies changed. A number of factors seem responsible for this. Pro-market Conservative governments were eager to involve the private sector in urban policy. Some central-government politicians were particularly interested in trying to expand the business dimension within urban policy. For example, Michael Heseltine, when Secretary of State for the Environment in the early 1980s, took representatives of major financial institu-tions to Liverpool in an effort to stimulate corporate interest in the fates of the cities. There was widespread evidence from American cities, such as Baltimore and Pittsburgh, that the private sector could be attracted into development projects within the cities given the right financial and political environments. It was also becoming increasingly apparent to business interests that continued urban decline would be an expensive process: physical and human resources would be wasted; urban civil unrest would increase; and corporate profits would decline (Confederation of British Industry, 1988).

These motivating forces have undeniably encouraged the private sector to take a greater interest in, and to devise programmes appropriate for, the older conurbations. However, any classification of these initiatives creates problems. There is, for example, the problem that a great deal of private-sector investment in the cities occurs through instruments that are devised and implemented by the public sector. This is true, for instance, of Urban Development Corporations and urban development grant, which are discussed in Chapter 6. In the remainder of this chapter, therefore, the emphasis is on programmes that have emanated from either the private sector or the private sector in conjunction with one or more other agencies, and that are at least in part relevant to inner-urban policy. Using this rather all-embracing definition, five approaches can be outlined: management buy-outs; secondments; business support organizations; Business in the Community; and partnership and development projects.

Management Buy-Outs

Whether management buy-outs should figure in an analysis of urban regeneration is debatable, but they are important in some conurbations in that successful buy-outs may retain output and employment while at the same time ensuring local control. Management buy-outs involve local managers acquiring financial support to enable them to purchase assets, sometimes from the receiver or from a retiring owner, but often from what may be an ailing parent company. By 1987, about 320 buy-outs were taking place, with the sellers receiving more than £3 billion in the form of loans from financial institutions that have been increasingly more willing to lend resources. A large proportion of buy-outs are for more than £25 million. Evidence suggests that many buy-outs do well and, after some initial problems, tend to increase profits and jobs (Wright and Coyne, 1985). Unfortunately, a large number of buy-outs occur in the more prosperous regions, notably the South East (Wright, Coyne and Lockley, 1984). There is undoubtedly a considerable scope for the approach to be encouraged further by local and central government and by business federations.

Secondments

Secondments are one of the more longer-standing private-sector initiatives. They are designed to provide such organizations as community groups, enterprise agencies (which are outlined below) and some government-inspired urban programmes (such as Task Forces, see Chapter 4) with managerial expertise. Some companies second staff directly, others use brokering agencies, such as the Action Resource Centre. Many urban programmes and business development organizations have benefited from appropriate secondments, particularly where the secondee has skills in areas such as marketing. On the other hand, secondees may not be fully committed to projects, and not all secondees make a happy transition from corporate environments to urban projects.

Business Support Organizations

The 1980s witnessed a veritable flowering of initiatives designed to support the formation and development of companies. In 1988, a review of these innovations pointed out that a wide range of supportive organizations able to provide assistance in such areas as technology, marketing and business practice, were in existence by the mid-to-late 1980s (DoE, 1988b). Some of these, notably science parks, owe their existence more to public-sector intervention than to private initiative (see Chapter 7). However, many other business support agencies, notably local enterprise agencies, came primarily from the private sector.

The enterprise agencies' origins are usually regarded as emanating with the St Helens Trust, founded in 1978 by the Pilkington glass company as a result of job losses in the industry. The evolution of that project has been well documented, not least because the policies and attitudes adopted at St Helens were to colour the evolution of the wider enterprise-agency movement (Hamilton Fazey, 1987). The St Helens Trust decided to adopt an approach that was based on the formation of a small, flexible, supportive agency, staffed largely by secondees. It obtained its resources mainly from local businesses and through some consultancy services. This type of approach tends to typify many of the 300 or so agencies established by 1988, although there are variations. Generally, they provide counselling services for small business, they run business clubs, provide linkages to financial and other networks, undertake courses for managers and those wishing to start up small firms, and they may have small-loan funds of their own. They have been funded in a variety of ways. Central goverment has helped through urban development grants, through Department of Trade and Industry support and through a specific Local Enterprise Agency Grant for smaller projects able to lever out equivalent private-sector resources. In addition, local authorities have helped and the private sector has been a major supporter with more than 4,000 sponsors by 1987. This has been encouraged in part by the tax advantages attendant upon contributions to enterprise agencies.

Enterprise agencies have been subjected to a substantial degree of evaluation (Deloitte Haskins and Sells, 1984; Centre for Employment Initiatives, 1985; Morison, 1987; Business in the Community, 1988; Segal Quince Wicksteed, 1988). There are positive aspects to them. There is evidence that they do help to create and maintain jobs in local communities. They may be as successful as other approaches in helping companies to survive the difficulties of early development (Thomas, 1988). They are cheap to run, they appear to have filled a management gap and they can act as a focus for local economic development in general.

On the other hand, there are issues that raise doubts about their future direction. There may be too many small agencies. There is a case for some rationalization of effort between agencies and amongst agencies, local authorities and the Department of Trade and Industry's Small Firms Service – all of which may undertake similar functions. There is also a case for greater specialization in urban areas, with different supportive bodies linked together through networking systems. Agencies may need to be more pro-active, seeking out clients and functions beyond their more traditional role. Importantly, the agencies' activities will depend on two other factors: on the micro scale, the role and performance of enterprise agency managers is vital; and, on the macro scale business support agencies operate in the wider economy. They can, along with virtually every other urban initiative, only change things marginally.

Business in the Community

Although many of the private-sector programmes discussed in this chapter emanate from – or have close connections with – Business in the Community (BiC), this organization merits specific comment because of its central role in stimulating private-sector investment in the cities and other economically-depressed areas. It was established in 1981, partly as a result of an Anglo-American conference on community involvement held the previous year. Its early evolution was relatively slow, but from the mid-1980s onwards (after successive Conservative electoral victories) the growing role of business interests in helping to devise and to effect such projects as Urban Development Corporations, and BiC's ability to attract royal patronage, meant that it became a much more significant agency in urban regeneration.

By the end of the 1980s, BiC was involved in a number of schemes designed in some way to enhance market investment in the older conurbations. Some of these have been mentioned already. It has, for example, encouraged the secondment of personnel from major corporations to community-based projects, and the formation of enterprise agencies both of which are discussed above. Some other aspects of BiC's work should also be mentioned (BiC, 1987; *The Financial Times*, 1987). It has, for example, entered into partnership with a number of other organizations to help create a co-ordinated approach towards aspects of urban renewal. These agencies include the Groundwork Foundation (funded by central government and the private sector, and designed to improve vacant and derelict land) and a number of schemes such as Project Fullemploy, Livewire, Instant Muscle and the Prince's Youth Business Trust, which are intended to provide assistance in the training of – or business support for – young people, particularly from disadvantaged groups. In addition, BiC has helped to foster the Per Cent Club through which, by 1988, over 150 companies had agreed to provide either 1 per cent of gross dividends or ½ per cent of UK pre-tax profits to community or charitable activities, job-creation schemes or inner-city initiatives. Finally, BiC members have undertaken individually a number of programmes designed to boost the economic fortunes of the inner areas. These include inner-city commercial, residential and industrial development projects, providing advice and assistance to new firms, employing and training disadvantaged groups, investment in urban science parks, purchasing and subcontracting policies designed to boost the output and profitability of small inner-city companies and direct financial support to community development projects and charities.

Partnership and Development Projects

Towards the end of the 1980s, one of the major trends in private-sector interest in the cities was the increasing involvement of the market in various projects designed to stimulate physical and economic development. Many of these were initiated in conjunction with other organizations, such as local authorities and community groups. Some of these schemes have been promoted by BiC (as was mentioned earlier), but the scale of private-sector involvement in urban development projects from about 1985 onwards justifies separate classification.

Some indication of the range of development activities undertaken by the market ought to be given. In 1982, for example, in response to pleadings by the Secretary of State for the Environment, Michael Heseltine, the major financial institutions created Inner City Enterprises, a development brokerage agency dedicated to urban-renewal projects based on private-sector resources and public subsidy. In 1988, British Urban Development (BUD) was established by eleven major civil-engineering companies. BUD has been pledged an initial authorized share capital of over £50 million and is designed to take a longer-term view of urban-renewal projects. Although it is intended that it will make a profit, it will be prepared to countenance initial losses on some schemes, and even cross-subsidization of less profitable by more profitable projects.

One of the key characteristics of many local development projects supported by the private sector has been the notion of partnership. In the context of urban policy, the concept of 'partnership' can mean many things, but in this case it is nothing to do with Partnerships discussed in Chapters 3 and 4. Rather it reflects the convergence of views and attitudes towards urban regeneration that have tended to epitomize relationships since 1985 between, on the one hand, business interests and, on the other, the overwhelmingly Labour-controlled cities. In the early 1980s, local government and corporate interests in a number of the cities were totally opposed. However, political and economic circumstances eventually imposed great changes on the attitudes of both parties. By the end of the 1980s, 'partnerships' were becoming commonplace. Glasgow saw one of the earliest of these, Glasgow Action, which consisted of a group of leading businessmen and politicians who, among other policies, tried to market the city as a centre of art and culture. In its attempt to hasten the re-orientation of the conurbation away from manufacturing and towards service-sector jobs, it can be argued that it met with a degree of success.

A more formal approach towards partnership was made in 1986, when the Department of the Environment helped to create Phoenix – a non-profit-making organization led by businessmen that was intended to act as a catalyst by bringing together private- and public-sector expertise to implement specific projects. Manchester and Bristol were two of the cities prepared to welcome the approach in its early stages. Other public–private sector development schemes and/or companies have been created in Leeds, Sheffield and Birmingham (Law, 1988).

Finally, mention should be made of possibly the most interesting of the development projects sponsored by the private sector: the Confederation of British Industry's Task Force on Business and Urban Regeneration. This initiative was designed to secure greater business involvement in aspects of urban regeneration in co-operation with local government and other bodies. It was intended that the approach should adopt a strategic vision by defining opportunities for private-sector investment in cities, and by isolating factors apparently responsible for failure. As part of its overall approach, the Task Force established 'The Newcastle Initiative', which brought together key businessmen and academics in the North East who ultimately argued for the implementation of key, flagship developments in Newcastle. These were based on policies designed to highlight the city's regional and cultural attributes, and its potential for enhancing links with Japanese investors.

However, the most significant development emerging from the Task Force was perhaps its publication, in 1988, of possibly the most substantial statement on the urban issue to be produced by the private sector. *Initiatives Beyond Charity* (Confederation of British Industry, 1988) makes very interesting reading. Whilst the approach predictably envisages large-scale private-sector investment in the cities stimulated by freer land markets and a simplified planning system, it also argues, among other things, for a partnership between public and private sectors, effective local-authority delivery in such areas as education and social services, more support for training initiatives and a boost to community enterprises. The obvious issue is not avoided – this will cost a great deal of money. The regeneration of 2,000 acres in inner Birmingham alone, it is estimated, will cost £1 billion. Some will come from private investment but much of the infrastructure and appropriate development subsidies will need to be funded by the State.

The Private Sector and the Cities: an Overall Evaluation

The incorporation of the private sector into aspects of urban renewal seems bound to remain a major feature of inner-city intervention into the 1990s. There are advantages in this. The resources available to the financial institutions alone are enormous – tens of billions of pounds. A great deal of this will not necessarily be used in the cities, but some will go into urban development projects. In addition, the private sector, through the Per Cent Club, through direct contributions, secondments and other means, will continue to finance specific projects within the cities.

The private sector will not just provide resources for the urban dimension. Expertise in areas such as marketing and management can be invaluable to smaller- and medium-sized firms. The policies of major companies in relation to purchasing, hiring and training can also play a significant role in providing opportunities for indigenous firms and disadvantaged groups. In the growing number of partnerships established throughout the country, the private sector will no doubt bring flair, imagination and business acumen to the proceedings.

The undoubted attributes the private sector will bring to urban regeneration should not, nevertheless, be allowed to hide some of the problems and dilemmas raised by this process. For example, the private sector is largely interested in profit. Some major corporations have extended their charitable and community activities, but the overall effect of these is not going to alter the fact that the private sector's dictates are dominated by profit. It was this dictate that did so much in the first place to undermine many of the economies of the northern cities through unfettered rationalizations and mergers, with the associated disappearance of so many indigenous firms in the late 1970s and 1980s – processes outlined in Chapter 2. There would inevitably have been a loss in manufacturing jobs, but the dramatic scale of employment and output loss in that period was unnecessarily intense and it was driven by market forces. It seems ironic that the market should be so eager to repair the damage ten years later, when the effects of its earlier actions have become apparent. It may be unrealistic to assume that the market would ever have imposed a self-denying ordinance on itself, but central government could have done much more through market controls to retain a manufacturing base both nationally and regionally.

Even if there is to be more private-sector interest in the cities, such intervention may well turn out to be rather specific in nature. Detailed evaluations of projects designed to attract the private sector into inner cites have highlighted predictably the market's unwillingness to invest in riskier projects, whose implementation will thus necessitate public-sector activity (Hart, 1984). This has always been apparent in large-scale industrial and commercial investment. Although there was some evidence of change towards the end of the 1980s, the financial institutions have always been reluctant to invest in industrial and commercial developments outside certain favoured locations.

This spatial patterning of market investment has encouraged the initiation and evolution of a number of public or quasi-public organizations designed to invest in areas that lack market appeal. This has been the case, to some extent, with the Scottish Development Agency (discussed in Chapter 6), English Estates (the government's industrial and commercial building agency, which has often proved the major initiator of projects in depressed regions) and organizations such as British Steel Corporation (Industry) Ltd and British Coal (Enterprises) Ltd (Hudson and Sadler, 1987), created to help redevelop older industrial areas – a function the market is unlikely to undertake. In short, the private sector will not invest in riskier projects or in certain regions unless there is substantial public-sector support, a constraint business organizations openly accept (Confederation of British Industry, 1988) even if central government sometimes purports to believe the opposite.

A number of other constraints may inhibit private-sector investment in the cities. Many business leaders are London based, and they show little interest in provincial cities. There also remains a legacy of political conflict between many Labour councils and the business sector. Finally, the business community can hardly have been encouraged in some of its early efforts in the cities. In Bristol, Manchester and Sheffield, for example, the local business community came into partnership agreements with local councils in relation to urban regeneration only to find that central government imposed an Urban Development Corporation.

It is important in this context to consider not only the potential shortcomings of private-sector investment but also its likely outcomes. A great deal of private-sector interest in the cities has been channelled into implementing specific projects in association with central or local authorities or both. The approach is often described as one based on the carrying out of 'flagship' projects. The private sector identifies market opportunities and the public sector provides support on a project-by-project basis. This model of urban regeneration has been described as a 'modified market' approach (Moore and Pierre, 1988), to be distinguished from both a pure-market strategy and what might be termed a socially-re-distributive model in which enhanced public-sector support is intended to provide opportunities for the disadvantaged through training, and support for companies. This latter approach was apparent in the early evolution of the Labour government's urban policy in 1977.

For many local authorities, political and economic realities mean that they have no alternative but to pursue this modified market approach. However, its implications and outcomes should not be ignored. It is, for instance, a notable example of how urban regeneration should not be administered. It depends on the implementation of a few *ad hoc* projects. Strategy is largely ignored. The renewal of London's Docklands (discussed in Chapter 6) indicates all too well the kinds of infrastructural problems unplanned redevelopment creates. Projects need to be located in broader policy. More important than this, however, are the restructuring implications for regeneration led by the private sector. Market strategies towards the cities tend to assume that it is the implementation of commercial, retail or cultural projects, or the building of high-cost residential developments in inner cities, that will somehow lead to a trickle–down of benefits to the less well-off – that development in effect creates jobs for everyone. There is little evidence for this (Boyle, 1988). Private-led strategies tend to be élitist in that they benefit the better-off, and by their very nature seek out opportunities that are most attractive to the market. The areas and people who need the greatest support receive least. Private-sector renewal may be a positive dis-benefit to many disadvantaged since developments tend only too often to involve the displacement of industrial land and buildings by commercial, retail or cultural activities that employ fewer people overall, and, particularly, fewer of the disadvantaged. This is not to argue that cities should be ignoring opportunities in such areas as tourism, leisure, commercial development and retailing. Intervention in these areas needs to be paralleled by other programmes designed to widen economic opportunities for that substantial group in society who are not going to benefit greatly from market-led approaches towards the cities.

Towards the end of the 1980s, it appeared that unemployment was beginning to fall in many cities. This was due to national economic performance: falling rates did not vary according to the urban strategies adopted by different cities. At the time of writing, it seems far from certain that the alarmingly-high unemployment figures will continue to decline.

Nevertheless, many urban authorities will have to accept the increasing role of the private sector within the urban dimension. They are scarcely in a position to refuse. They have no alternative resources, and co-operation with the market will enhance development. However, this drift towards development projects emerging from increasing public–private-sector co-operation should not be equated with an urban strategy. It remains a pragmatic, essentially *ad hoc*, very partial approach to the range of problems faced by the cities. It does nothing for urban services, for the financial crises endured by many councils, and very little for unemployment.For many of the disadvantaged living in the inner cities and on peripheral housing projects, the urban dimension becomes a greater irrelevance at a time when central government continues to withdraw from a range of welfare functions. For the disadvantaged, some of the best opportunities may well lie with community enterprise. However, as earlier discussions here have indicated this is not a particularly appealing prospect in many parts of the country.

PART III
Inner-Urban Policy: Evaluation, Explanation and Reform

PART III

9
Inner-Urban Policy: Overview and Critique

Part II attempts to explore some of the major thrusts of inner-urban policy in the UK. Some evaluation of particular initiatives is developed, but there is a need to present a more integrated critique of inner-city intervention as a whole. This chapter aims to achieve that objective in two ways. The bulk of the chapter is dedicated to a critique of the operation of inner-city policy. Although reference is made where appropriate to initiatives emanating from sources other than central government, the emphasis is inevitably placed on what the centre has implemented – that, after all, is what is generally understood by inner-urban policy. Later in the chapter, brief reference is made to the outcomes of inner-urban policy – its effects on the cities. The problem with this task, important as it may seem, is that it is extremely difficult to measure the direct effects of a programme as marginal as inner-urban policy. However, its insignificance in relation to the state of the cities at the end of the 1980s ought to be established.

Inner-Urban Policy: Organization, Structure and Operation

Attempts to devise an evaluative framework for inner-urban policy have been assisted by the contributions of a number of observers (Barrett and Fudge, 1981; Stewart and Underwood, 1982; Barrett and Hill, 1984; Hambleton, 1986; Solesbury, 1986, 1987; Stewart, 1987). Whilst these commentators present divergent views on the origins and development of inner-urban policy, one theme that tends to emerge is the need to evaluate such policy areas as the inner-city programme within a hierarchical framework. Effective policy analysis needs to locate developments within a complex interweaving of processes operating at different levels. On one level, a series of considerations relating to ideological issues merits examination. Debates about the role of the State and central–local relations must be considered if inner-urban policy is to be understood fully. These types of ideological debates are fundamental in explaining the origin and development of inner-urban policy, and are considered in Chapter 10. In this chapter, where the focus of attention is directed towards an overall evaluation rather than explanation of policy, attention is directed towards two lower levels of evaluation: institutional and implementational.

Institutional Considerations

Inner-urban policy has emerged from a series of ideological concepts, notably those pursued by anti-collectivist governments elected after 1979. Nevertheless, despite the impact of these broader constraints (discussed in Chapter 10), a number of what might be termed institutional factors need to be explored if the policy area is to be understood and effectively evaluated. Three issues seem relevant: the relationship between analysis and prescription, the question of co-ordination and objective definition.

Analysis and prescription

Inner-urban problems have been subjected to very substantial analysis. As the most cursory glance at any appropriate bibliography will reveal, inner-urban problems have produced an extraordinary outpouring of publications from central and local government, trade-union and employer organizations, churches, charities, academics and community groups – one thing the inner cities are not short of is analysis. The quality of analytical research varies enormously, but, in particular, substantial and comprehensive reviews of economic and employment problems have recently been forthcoming (Hausner, 1986; 1987). While an overall assessment of these findings would be inappropriate, one or two major conclusions can be explored in order to illuminate the marked dysfunctions between, on the one hand, analysis and, on the other, policy development.

One central conclusion that would emerge from any review of the inner cities is the sheer scale of the problem. Between 1971 and 1981, the inner cities of the seven major conurbations lost half a million jobs (Begg, Moore and Rhodes, 1986). By 1987 the travel-to-work areas of these conurbations accommodated more than 850,000 unemployed people and, with the exception of Greater London, unemployment rates were officially everywhere over 15 per cent, and unofficially much higher. In addition, estimates of expenditure needed in such areas as urban infrastructure (Cowie, Harlow and Emerson, 1984) and housing repair (Association of Metropolitan Authorities, 1986), run to tens of billions of pounds. Whatever else the inner-city problem might or might not be, it is not going to be moderated without substantial across-the-board investment.

Not that this would become apparent from a review of the resources allocated to inner-urban intervention – the marginal scale of these has been alluded to frequently in this book, but it merits a brief comment here. Total funding is difficult to estimate because of the diffuse nature of the whole initiative. However, the government has argued that the Department of the Environment directed about £3,350 million to the inner areas between 1979 and 1987 (Patten, 1987). In addition, MSC funding, specifically for the inner areas, amounted to about £100 million in 1985–6, and smaller sums collectively worth about £50 million were allocated by the Department of Trade and Industry through its innovative and small-firms schemes, and by the Home Office on support for ethnic-minority projects (House of Commons Committee of Public Accounts, 1986b).

The government has argued that total expenditure on the inner cities from all appropriate budgets amounts to £3 billion per annum (HMSO, 1988a). Even if that were true (and in practice this figure undoubtedly includes expenditure that would in any case have been spent on the cities irrespective of any specifically-urban bias), it still seems insignificant next, to say, the £170 billion per annum of total public spending, or the £50 billion the Confederation of British Industry has argued ought to be spent on the cities (Confederation of British Industry, 1988). The scale of the problem highlighted by analytical observation is not matched by the resources allocated to the urban dimension. This discrepancy between analysis and prescription can be seen in many aspects of inner-urban policy, but two other instances might be discussed profitably: the degree to which a specifically 'inner'-urban problem can be identified and the relationship between regional and urban issues.

The proliferation of inner-urban initiatives is based on the assumption that a specifically inner-urban problem can be identified – that conditions are significantly worse within the inner cores of Britain's cities than elsewhere. There is clearly some evidence for this (Begg, Moore and Rhodes, 1986). The inner-urban areas have lost populations faster than the outer-urban zones, and they have higher unemployment rates. This might be regarded, therefore, as a justification for an inner-urban policy.

Several points ought to be made here. First, the causes of inner-urban decline, such as the decentralization of the more wealthy or the restructuring of production (see Chapter 2), will not be addressed through policies that institute change in the inner cities. The causes of these processes lie elsewhere. Second, inner-urban problems cannot justify the spatial patterning of intervention apparent within Britain's inner-city programme. Many urban initiatives deal with relatively small areas, into which additional or re-allocated resources are directed. This approach might seem suitable to a government committed to limited intervention, but it fails to address the problems of the inner cities as a whole. Finally, a too-close concentration of activities in inner cities deflects attention away from other deprived areas and deprived people. Some outer-urban public estates may suffer from problems very similar to those apparent within the inner cores, and they merit interventionist support. Policies directed towards areas may have little impact on deprived households, most of whom do not live in small, defined, inner-urban cores of the largest cities.

One final indication of the discrepancy between analysis and prescription should be outlined: the lack of integration between regional and urban policy. There is evidence that 'efforts to improve the performance of urban economies must go hand in hand with measures to improve the performance of the regional economy of which the urban area is part' (Wolman, 1987, p. 37). It would be difficult, however, to argue that there has been, in practice, anything but the most fleeting of relationships between urban intervention and regional policy. Regional policy, in particular, has been steadily and substantially curtailed through out the 1980s, such that resources of £360 million allocated to it in 1985–6 amounted in real terms to about a half of its mid-1970s figure. With the 1988 decision to abandon automatic regional development grant, it would appear that the cities cannot expect central government to provide a realistic system of regional support. As with so many aspects of inner-urban intervention, prescription bears little relationship to analysis.

Co-ordination

For many years, co-ordination has featured as a prominent objective of inner-urban policy. The 1977 white paper identified the need for a unified approach to the inner areas (HMSO, 1977). This sentiment has been echoed by succeeding Conservative governments (DoE and DE, 1987). The primary focus of some urban initiatives launched in the 1980s – notably the City Action Teams and Task Forces discussed in Chapter 4 – has been to achieve a co-ordinated governmental response to urban problems and policies. *Action for Cities*, the 1988 statement on the urban cores, made it clear in turn that the main thrust of the government's programme was designed to co-ordinate existing approaches and not to intro-duce new initiatives or new resources (HMSO, 1988a).

Whatever the argument, in practice, as the Town and Country Planning Association states, far from there being a co-ordinated approach, 'there has been the opposite; a series of piecemeal measures which do not add up to a co-ordinated policy' (Town and Country Planning Association, 1986, p. 9). A genuinely co-ordinated response to the problems of the cities would require an integrated programme from virtually every government department. Even depart-ments such as Defence, which has allocated so much of its procurement spending to the South East and South West of England (Lovering and Boddy, 1988), would need to be absorbed into a cohesive package of investment and support for the cities.

In practice, the probability of achieving this objective looks extremely slight. So many organizations – not just central-government departments – are inter-ested in, and have resources for, the inner cities that attempting to devise a consensual strategy incorporating all interested parties appears impossibly diffi-cult to attain. Agencies tend to retain different objectives, are sustained by contrasting political forces and adopt varied operational structures. It may even be difficult to synthesise a coherent urban strategy in one conurbation, because of the markedly-contrasting attitudes and objectives of relevant local bodies (House of Commons Environment Committee, 1983). It appears somewhat surprising that Conservative governments continue to pursue the idea that co-ordination is attainable when the most successful developmental initiative has proved to be the Urban Development Corporations – the least collaboratively inclined of all inner-city programmes.

One of the assumptions inherent in debates about co-ordination is that it is an intrinsically-desirable objective. Is it? It assumes that from the myriad of agencies that might have a bearing on the fate of the inner cities, some overall and implementable strategy can be found. This assumption appears extremely naïve. It is apparent that, even in central government, some departments (notably Trade and Industry) have traditionally been less than enthusiastic about an inner-urban programme (Hambleton, 1986).

This type of conflict is likely to be mirrored in turn at the local level. Business and employer organizations have different views on land use, the role of local government, the need for infrastructural investment and local labour-market policies, than will community groups or many local authorities. In local government itself, different departments have contrasting views on longer-term strategy, as any observer of the debates surrounding appropriate inner-city land-use plans will be aware. Conflict not consensus dominates issues as contentious as inner-urban policy. The resolution of this conflict lies either with traditional democratic processes, where local government – as the elected organization – undertakes its own debates and emerges with a defined programme in consultation with central government, or a development agency is imposed. Evidence from the Urban Programme, from the Partnerships onwards, shows that co-ordination is not the answer.

Objective definition

Problems of co-ordination highlight and reflect the allied issue of objective definition. With so many organizations involved in some aspect of inner-urban policy, it is to be expected that a clear and definitive statement of goals has been illusive. Guidelines published by the Department of the Environment in 1985 suggest that the Urban Programme should attempt to achieve four major objectives: improving employment prospects, reducing derelict land, strengthening the social fabric of inner cities and reducing housing stress (DoE, 1985c). These objectives appear largely non-controversial, but they do not address two important problems: the inter-related nature of the urban problem and conflicts in objective definition.

The inter-related nature of urban problems has been stressed by many observers. To take but one of many examples, the Scarman Report argues that the disorders in Brixton cannot be understood unless 'they are seen in the context of . . . complex political, social and economic factors' (Scarman, 1981, p. 15). The complex and multi-faceted nature of urban problems requires the formulation of appropriate policy objectives, but this tends not to occur. For example, the standard objective of improving employment prospects, which tends to figure in virtually all inner-city programmes, raises a whole series of subsidiary debates relating, for example, to the kinds of jobs that ought to be sought out, for whom and how. Equally importantly, the relationships between, on the one hand, employment objectives and, on the other hand, investment in training, education, transport and housing, tend to be played down. Yet, as Chapter 7 shows, the best way of creating jobs might be to invest in such areas of consumption as housing and social services – a strategy unlikely to be adopted by Conservative governments determined to reduce public expenditure.

There is also the parallel constraint of conflict between objectives. This tends to be particularly apparent in inner-urban policy in that a more or less permanent tension has existed between what might be termed welfare objectives and developmental aims (Solesbury, 1986). Welfare objectives, which have a long tradition in British inner-urban policy, are primarily concerned with such issues as policing, training, education, poverty, housing and community and co-operative ventures. Developmental aims come from rather different sources. They tend to be rooted in supply-side considerations, and they see urban policies as needing to be framed around such issues as financial and physical inducements to markets, deregulation, infrastructral investment, and so on. These two sets of objectives need not conflict. For example, the early evolution of some enterprise boards was structured around the notion that public-sector intervention in such areas as equity investment and property development, could proceed, while at the same time the jobs so created might be allocated, where possible, to the disadvantaged. Even if developmental and welfare objectives need not be mutually incompatible, the 1980s have seen, in practice, a steady weakening of welfare objectives to the benefit of developmental aims. Issues of poverty and deprivation have tended to be eliminated from the inner-urban debate.

Policy Implementation

The pragmatic nature of intervention

One consistent criticism of inner-urban policy as a whole has been its *ad hoc*, pragmatic nature. There has been little or no attempt to devise a strategic view of what should happen to the cities as a whole or to individual conurbations. Broader policy issues seem to have been increasingly avoided by central government. This lack of strategy has been commented on, amongst others, by those examining land acquisition policies for economic development projects funded under the Urban Programme (JURUE, 1986b), the House of Commons Environment Committee in its consideration of urban initiatives in Merseyside (House of Commons Environment Committee, 1983) and a House of Commons Committee of Public Accounts report into the Urban Programme (House of Commons Committee of Public Accounts, 1986b).

There seems to be a number of factors that might account for this lack of strategic vision. Problems associated with co-ordination and objective definition, referred to earlier, also seem pertinent to this debate. Bearing in mind the problems of integrating policy objectives from the range of central-government departments and organizations that would want some say in the creation of any strategic urban vision, it is perhaps naïve to assume that a coherent strategic programme could ever emerge from central government. Moreover, it is apparent that many urban projects developed by central government in the 1980s, have been specifically intended to pursue *ad hoc* projects without attempting to locate these within a broader vision of what should be happening to conurbations as a whole, or to places, classes or employment sectors within conurbations. Parkinson and Duffy's (1984, p. 87) comment that the Merseyside Task Force hardly began 'to grasp the nettle of strategic analysis', could apply equally well to many other urban schemes initiated in the 1980s.

There is also the point that in many inner-city schemes, administrative procedures have come to dominate activity (Stewart, 1983). This tendency has been seen (Hambleton, 1981) as a shift away from policy, with its concern for analysis, negotiation and innovation, towards a programme approach sustained by fixed budgets and arbitrary time-scales that inevitably reduce the role of strategic debate and policy development. This has occurred, for example, in the Urban Programme, where the Urban Programme Management Initiative (UPMI) imposes formidable managerial tasks relating to quantitative assessments of specific projects on local authorities, but discourages strategic evaluation.

Innovation and the urban dimension

It had always been anticipated that the urban dimension would be innovative – the 1977 white paper, for example, foresaw considerable changes being implemented in such fields as housing management, the delivery of social services and local economic development (HMSO, 1977). These sentiments were reiterated some years later when the Urban Programme was seen as being 'well suited to support pilot and experimental projects' (DoE, 1985c, p. 28). Inevitably, with so many initiatives emanating from the entire urban dimension, some innovative programmes have emerged. This is particularly true of those projects implemented either by local government or by community development, which are discussed in Chapters 7 and 8 respectively.

Alternatively, innovation has not been a particularly marked feature of the major inner-urban initiatives introduced by central government. This is particularly unfortunate since it is central government that still provides the largest resources for inner-urban intervention, and that acts as the hub around which the entire edifice revolves. Nevertheless, despite its important role in the urban dimension, many projects effected by central government appear to lack an innovative edge. To those evaluating the Birmingham Partnership, for example, it appears that 'many projects have been neither clearly distinctive in approach from main programmes nor genuinely innovative' (Aston University, 1985, p. 111). Others examining the Liverpool Partnership have come to much the same conclusion (Parkinson and Wilks, 1986). These sentiments have been echoed by other commentators who assess inner-city policy as a whole. To Stewart, for example, 'in general terms, inner cities policy has not been the vehicle by which innovation either in policy design or policy delivery has been supported' (1987, p. 137).

These assessments, of course, need to be put into context. Many observers perceive a lack of innovative practice specifically within the Urban Programme, and not in central government's inner-city dimension as a whole. Even if evaluation is widened to include other aspects of central government's urban dimension, the overall structure does not appear particularly imaginative. Urban development grant and principles of leverage had been practised for many years in America (Boyle, 1985). The concept of deregulation, central to enterprise zones, was not new to the international arena. Autonomous development agencies, the Urban Development Corporation model, had also been in operation in America for many years, and in Scotland since 1975. However one views inner-city policy, innovation is not a dominant characteristic of programmes initiated by central government. The innovation that has occurred has emanated overwhelmingly from local authorities and through community development.

This lack of innovation in much of central government's inner-city dimension can be related to a number of factors, which are discussed elsewhere. A lack of strategic vision seems relevant in that a broader framework would give central government, and also other bodies, the freedom and confidence to innovate within defined parameters. This also applies to the problem of departmentalism evident within both central and local government. Too many organizations have viewed urban funding as a means through which additional (if admittedly small) sums of money can be obtained to implement schemes that would have ultimately been funded through orthodox channels. The heavy emphasis on capital spending within the Urban Programme has not helped in this respect. Current, not capital, funding is more likely to encourage innovation. Finally, the vetting of Inner Area Programmes by local chambers of commerce and industry is probably going to inhibit rather than stimulate experimentation.

Monitoring and evaluation

From the inception of the inner-city dimension, central government has argued that, because of the ostensibly innovative nature of urban experimentation, projects should be monitored and appropriate findings disseminated. It is evident that this task has been consistently problematic. For a number of years, for example, the entire Urban Programme was largely unmonitored, until the introduction of the Urban Programme Management Initiative in 1985, which devised a standard set of numerical output measures against which to assess virtually every project (DoE, 1985c). Equally, from the mid-1980s onwards, the Department of the Environment commissioned a substantial programme of research into many aspects of urban intervention. Some of the resulting reports were little more than practice manuals (DoE, 1987b) or were based on relatively few case studies (JURUE, 1986a). Other publications reflected altogether more substantial pieces of work (P.A. Consultants, 1987). Nevertheless, despite a generally more progressive programme of research that began to appear from the mid-1980s on, problems of evaluation and monitoring have remained.

One issue that has particularly hindered the Urban Programme is that too much of the monitoring tends to be overly detailed in nature. Financial budgets, time-scales and simple quantitative assessments have dominated evaluation. Broader, and far more important issues governing the extent to which a particular policy appears to be achieving specific ends, tend to be neglected, perhaps an understandable deficiency when all too often objectives are unclear, meaninglessly over-generalized or likely to conflict with other stated goals. Evaluation should lead logically to policy modification. Inappropriate programmes should be abandoned and more effective schemes promoted. Yet there is very little evidence to indicate that monitoring of inner-urban initiatives has led to many substantial modifications in policy – a criticism that could be levelled at many other strands of public policy.

It seems surprising that the Urban Programme (generally regarded as an over-bureaucratic irrelevance) should have remained in existence for so long. If creating jobs is seen as a central goal of inner-city intervention, there can be little doubt that local government has proved altogether more cost-effective than has central government. Far from encouraging local innovation, Conservative governments elected in the 1980s have generally tried to subdue council initiatives (see Chapter 7). Monitoring inner-city intervention, which in some respects has been substantially enhanced since 1985, has not resulted in the scale of policy review which might have been expected from its findings.

Although the most significant fault in monitoring inner-urban policy has been the inability of evaluative studies to modify aspects of the urban dimension substantially, it is apparent that genuinely-complex methodological problems surround many monitoring exercises. These have been particularly noticeable in areas of employment creation, and they have received considerable comment (see, for example, Gregory and Martin, 1988; Horne, 1988; Monck *et al.*, 1988; Turok, 1988). This debate cannot be addressed in detail here, but some indication of the complex interplay between different factors should be outlined. If, for example, a monitoring exercise is attempting to discover the cost-effectiveness of a particular type of initiative designed to enhance economic output and/or employment generation, a daunting list of variables may need to be produced. It may be relatively easy, for example, to establish the gross numbers of jobs associated with a particular project, but this raw figure will need to be refined according to a number of processes – some of which will tend to increase it and others to reduce it.

On the positive side, public-sector intervention that culminates in, say, the creation of a new science park, or enhanced output from firms supported by enterprise board equity investment, will have a number of additional benefits to the local economy concerned. In particular, firms so assisted may exert multiplier effects in that more local goods and services are acquired than would otherwise have been the case, thus boosting output and possibly employment in other companies. Equally, public support for enterprises that develop or utilize new products and processes may demonstrate to other companies the benefits of innovative approaches, although the degree to which demonstration effects operate may be very difficult to gauge.

Alternatively, there are a number of processes that will tend to diminish the real effect of public support in the fields of employment generation and economic development. For instance, the results of public-sector support through programmes such as municipal purchasing of local goods and services, or equity investment in companies, or providing new industrial property, is too often taken as the gross job total. In other words, the numbers of people working in projects is collated, and this is presented as if all such jobs were dependent on public support. These totals are misleading: they take no account of either displacement or additionality. Support for one company or project may simply displace activity elsewhere. Crucially too, it is often assumed that support leads to additional employment and output. In reality public subsidy may be unnecessary – much the same would have occurred had the subsidy not been forthcoming.

Assessing the degree of additional activity caused by support is clearly far from easy, but it might be attempted by using such techniques as asking managers what new output and employment resulted from subsidy, or evaluating activity in supported projects in relation to those that did not receive support. The methodological problems associated with gauging the real impact of public subsidy are formidable, but if working assessments of the effectiveness of specific programmes are to be attempted, this kind of exercise is vital. Finally, and to reiterate themes developed in Chapter 7, where efforts are made to explore the net costs of job creation through time (as jobs created through subsidy will often last for more than one year) local-authority financial assistance is capable of creating jobs at figures that equated roughly with the annual net costs to the Exchequer per person unemployed – in other words, nothing (Turok, 1988). Equivalent figures for such schemes as enterprise zones, and even regional policy, are dramatically higher than this. Far from attempting to enhance local-authority job-creation programmes, Conservative governments elected since 1979 have attempted to diminish them. This says little for effective monitoring.

Inner-City Intervention: Policy Outcomes

Many attempts to evaluate inner-urban policy have tended to be rather introspective – the emphasis has been on the operation and identifiable effects of specific schemes. Many of these issues are discussed in Chapters 3, 4, 5 and 6. What tends to receive far less attention is the effect inner-urban intervention as a whole has had on the major conurbations. It is perhaps not surprising that this task has not figured prominently in many assessments: it is difficult to highlight changes in cities because of what is, as this book often points out, a very marginal policy.

Nevertheless, a brief if somewhat impressionistic overview of the effect of inner-city policy as a whole should be attempted.

Although a central theme of this book is that urban policy has been very inadequate in dealing with the problems of Britain's cities, it is apparent that certain aspects of the programme have been successful in terms of defined objectives. The most noticeable instance of this is property development. What Urban Development Corporations, urban development grant and enterprise zones have shown is that, given sufficient subsidy, private-sector property development will occur in many parts of urban Britain. The one notable success in the Conservative's approach to the inner areas is physical development, not just in London's Docklands but also in many other places subject to inner-urban assistance. However, this apparent success needs to be placed in context. A public subsidy is associated with physical development. Much of this subsidy may be deadweight – projects would have occurred with less or even no subsidy. Who is benefiting from this development? Not the poor, to any significant extent. If the policies designed by Conservative governments have resulted in greater private-sector interest in the cities, much of this interest is clearly fuelled by the profit motive, not philanthropy. Although inner-urban policy has helped to sustain property development in the cities, it is clear that, taken in its entirety, the programme has failed to meet the cities' needs. Contemporary urban *malaise* is examined in Chapter 2 and does not need reiteration here, but a number of particular failings in inner-city intervention should be highlighted.

The programme has proved to be a badly co-ordinated series of independent schemes that lack any coherence. It has not addressed in any effective manner a series of constraints that affect the older urban cores of Britain and the people living there. Different observers would highlight contrasting failures of inner-city intervention, but six issues merit particular comment: unemployment, poverty, race, policing, political participation and public services.

In 1987, over 800,000 registered unemployed lived in the seven major conurbations of the UK. Since methods of collating unemployment statistics have been changed almost twenty times since 1979, and almost all these changes have reduced the numbers of those eligible to register, the real figure is likely to be substantially higher. Yet unemployment has come to figure less and less prominently as an issue in urban policy. The emphasis has switched to development, especially property development, which, while enhancing employment opportunities for those in the cities during construction, is unlikely to employ many afterwards. Inner-city intervention has failed to relieve unemployment to the extent it should have and, as Chapter 11 points out, could have done.

If unemployment has not figured as prominently as it might have in inner-city intervention, the deprivation debate appears to have disappeared entirely from the agenda. Yet, as Chapter 1 shows, the origins of inner-city intervention lie with projects whose fundamental objective was relieving poverty. The community development projects and other studies may have been unable to moderate deprivation substantially, but they certainly highlighted its extent, causes and the policies that might alleviate it. Little of this features in contemporary urban debate, with its concern for development. Single-parent households, the elderly, the poor and the disabled, whose lives continue to be documented (Town and Country Planning Association, 1986; Benyon and Solomos, 1988) can expect little from urban policy in the late 1980s.

Another factor – race – was very pertinent to the early development of inner-city policy. The announcement of the Urban Programme in 1968 has often been seen as a direct response to the anti-immigration speeches of Enoch Powell and others. Yet the inner-city programme developed in the 1980s has made minimal reference to the fate of ethnic minorities in the urban cores. Some Partnership projects are intended to boost ethnic business and advice centres for these enterprises have emerged. But the scale of problems facing black people, which have been examined elsewhere (Gilroy, 1981–2; Broadwater Farm Inquiry, 1986; Benyon and Solomos, 1988), will not be diminished in any measurable way by inner-city intervention.

This conclusion applies to questions of policing in the cities, and to London in particular, which a whole range of commentators see, at the very least, as problematic (Scarman, 1981; Policy Studies Institute, 1985; Broadwater Farm Inquiry, 1986). Issues of harassment, racial discrimination by the police, the role of community policing and police accountability remain important in the urban debate. Inner-urban policy, however, has said little about them, partly because, while the Home Office retains a remit over policing, it plays a very small role in devising inner-city policy, and partly because of a lack of political will.

A fifth and often neglected factor in the urban debate is the question of political participation. Many households, groups and ethnic communities in the older urban cores are regarded as effectively excluded from orthodox political activity (Scarman, 1981). As Benyon and Solomos point out, 'opportunities for participation through established channels such as parties and pressure groups tend to be relatively infrequent in inner city areas' (1988, p. 417). Predictably, and bearing in mind the shortcomings mentioned above, inner-city policy has scarcely begun to explore mechanisms through which greater participation in political activities might be engendered. It could be argued that community groups have been incorporated into some projects, such as Partnerships, but the process is very limited. It may be that widening political participation is not an activity inner-city policy can do alone or at all, and that political parties should be more enterprising in this respect. However, it needs to be done. Feelings of political exclusion must be seen as a prime factor in explaining the urban disturbances of the 1980s.

Finally in this list of shortcomings, mention should be made of public services. Across the board, they have to be evaluated as totally inadequate. Estimates for improvements to the urban housing stock ran to £19 billion in 1985 (Association of Metropolitan Authorities, 1986). In some cities, sewerage and drainage systems are collapsing. Capital investment in social, educational, recreational and health services is desperately needed. Transport services, so often a neglected aspect, need substantial improvement. The extent of investment in public-transport systems in many European cities makes British efforts in this field tiny by comparison. Many conurbations that have been prepared to invest in new transport systems appear to have been more successful in regenerating their older urban areas than have been those remaining dependent on older and declining systems (Law *et al.*, 1988).

Other issues relating to policies that might be incorporated into a more effective inner-city programme merit further consideration, and are explored in Chapter 11.

10
Inner-Urban Policy: Towards an Explanation

Questions relating to an explanation of British inner-urban policy raise a number of complex issues. For example, explaining why an urban dimension emerged in the late 1960s may well invoke different kinds of debates from, say, discussions surrounding inner-city policy carried out by Conservative governments elected since 1979. More fundamentally there is the problem that inner-urban policy, as is developed below, can elicit 'explanations' from diverse theoretical and ideological sources. This is a constraint that affects other policy areas, but it is perhaps particularly evident in relation to inner-city policy. This is because the problem is so difficult to define (if a specifically inner-urban problem exists at all), and because policy initiative has been so diffuse, incorporating everything from job creation to neighbourhood crime-watch projects.

Despite these reservations, this chapter attempts to highlight some themes that might be used to explain or to locate inner-urban policy. It is not intended to provide a comprehensive review of all the theoretical developments that might in some way affect inner-city policy – that task would require an entire book of its own. Here, where the emphasis is placed throughout on an evaluation of inner-city initiatives, it is important that some key constructions are identified and briefly evaluated. Anything more comprehensive would be inappropriate.

This chapter consists of two major sections. The first part discusses explanations rooted in policy analysis. The second part develops ideologically-inspired constructions.

Policy Analysis and Inner-Urban Policy

A number of commentaries on inner-urban policy have evaluated its origins and development in what might broadly be defined as policy analysis. Observers have attempted to understand or to describe the evolution of urban intervention within prevailing administrative and political parameters. Some observers have left the debate there. In other words, policy is seen to have emerged from – and can be understood in the context of – the interaction between politicians, local- and central-government personnel, the business community and other factors and agencies. Other observers argue that, although these considerations are important, there remain underlying ideological issues that need to be addressed if an accurate understanding of inner-city policy is to be achieved. Some explanations therefore combine issues raised below with broader ideological themes raised later in this chapter.

In the 1960s and early 1970s, policy analysis – urban and regional planning being good examples – was based on assumptions of rationality. It was assumed that policy objectives could be readily identified, that a range of viable strategies could be highlighted and that a best solution could be selected. However, as a number of commentators point out, these ideas may have little bearing on reality (Wildavsky, 1979; Heclo and Wildavsky, 1981). It may be difficult to identify exactly what constitutes a problem area, or to define precise objectives for policy intervention. In many instances it may be impossible rationally to devise genuinely-different strategies to overcome defined problems and to be in any position to arrive at an unambiguously 'best' solution.

Many of these comments apply to inner-city policy: the problem itself is not easy to define; it can be seen as relating primarily to lost jobs, but other considerations, such as crime, poor housing, environmental degradation, declining services and poverty, will be regarded by such sections of the community as single-parent families, the elderly, the disabled, and so on, as vital to their understanding of 'the inner city'. Taking the argument a stage further, there are those who believe there is no inner-urban problem at all, that the economic processes affecting the cities are not spatial in their operation, and their outcome by way of lost jobs and investment can be seen in many places, not simply in the inner areas of the major conurbations.

If there are initial difficulties in defining what the inner-urban problem is, these become insignificant when compared to the problem of defining appropriate objectives and strategies. In the case of objectives, it may be impossible to decide rationally whether inner-urban policy should strive to retain populations or to allow them to decline; to accept or to try to reverse urban manufacturing decline; or to encourage or otherwise the 'greening of the cities'. Objectives for inner-urban policy are ultimately politically articulated, and there is no rational nor unequivocally correct list of goals. As much of this book tries to show, there is no definitive 'best' inner-urban strategy to adopt. Different commentators promote contrasting approaches based largely on ideological considerations. In short, the assumption that inner-urban policy can be subjected to rational policy analysis has to be rejected – approaches based on different assumptions need to be explored.

In practice, however, those commentators who address questions relating to the origin and development of inner-urban intervention from the point of view of policy analysis, have produced more sophisticated explanations than would emerge logically from investigations based on rationality (Edwards and Batley, 1978; Barrett and Fudge, 1981; Stewart and Underwood, 1982; Hambleton, 1986; Stewart, 1987). It would appear from these contributions that a number of themes could help to locate and explain the origin and development of inner-city policy.

Politics and Politicians

The evolution of inner-uban policy has been consistently influenced by politics. A fine line has to be drawn between ideological considerations discussed later in this chapter and the more immediate political actions that form the focus of attention here. Many political interventions are rooted in deeper ideological conviction, but not all of them. By the mid-to-late 1980s, one could argue that inner-city policy had become a party-political issue: the market-orientated approach pursued increasingly by Conservative governments would not have been that effected by a non-Conservative administration. Ideological issues were coming to dominate policy.

However, in earlier years (from the late 1960s to the late 1970s), there was less party-political antagonism. Instead, there was a widespread consensus that an inner-urban problem existed, and that resources should be allocated by central government to deal with the problem, working in association with local government and other agencies. In this earlier period, when the urban experiments (see Chapter 1) were launched, and when, between 1977 and 1979, more permanent policy was effected (see Chapter 3), politically-inspired interventions were extremely important in moulding the urban dimension. As Edwards and Batley (1978) point out, in the late 1960s, the very creation of an urban policy almost certainly reflected a determination on the part of Wilson's Labour government to do something to counteract anti-immigration speeches made by Powell and others. This reactive response was apparent when Callaghan's government, through Peter Shore, Secretary of State for the Environment, introduced permanent inner-urban policy. There may have been a logical argument for this step, but it had obvious political advantages: it directed resources to the cities (an important electoral consideration for a struggling Labour Party); it might help to create jobs when unemployment was rising; and votes were possibly to be had in moving resources away from the new towns that had little popular appeal.

This willingness to use inner-urban intervention is by no means confined to Labour administrations: Conservative governments have been equally prepared to use it, particularly some Secretaries of State for the Environment. For example, in the early 1970s, Peter Walker in Heath's government was fully prepared to heighten the profile of his department and, by implication, his own political standing by initiating a number of urban projects, notably the Inner Area Studies.

Perhaps a better example happened a decade later. Michael Heseltine was always an interventionist Secretary of State for the Environment, promoting, among other things, Urban Development Corporations and urban development grant. After the 1981 riots he proposed a series of policy developments, notably on Merseyside after well-publicized visits to Liverpool. Interestingly, some of his ideas, such as the creation of a Minister for the Cities, were never implemented. However, some projects were set up in Liverpool, and the financial institutions were encouraged to visit and to invest in the cities, a move that bore fruit later in the 1980s. A number of interpretations can be placed on Heseltine's activities after the riots, but there can be little doubt that his interventions at that time did him no political harm.

The best example of politically-inspired interventions in urban policy perhaps came in 1988, with the Conservative government's publication of *Action for Cities* (HMSO, 1988a). At the press conference launching the initiative, the Prime Minister candidly accepted that the approach contained no new policies and no new money – it was simply a reiteration of existing programmes with some minor modifications. Why was it published? Possibly because the government had to appear active in an area that had suddenly become a central political debate, to be superceded a few months later by the 'green' issue. While more profound explanations for urban intervention can be identified, the role of narrowly-political forces in guiding and structuring policy should not be ignored.

Organizations and Administrative Structures

Inner-urban policy emerges from, and is implemented by, various agencies: local authorities, quangos, several branches of central government, the voluntary sector, private companies, and so on. Inner-city policy is not alone in emanating from a range of factors and agencies, but few policy areas could claim to encompass so many bodies in such complex resource and administrative relationships. This complexity gives rise to a number of observations that might help in the understanding of the evolution and development of urban intervention.

One issue already mentioned on a number of occasions is the question of co-ordination. The cities' problems are multi-faceted and require assistance from many agencies. However, one of the most important constraints on inner-city intervention has been the inability of relevant agencies to achieve a degree of internal or external co-ordination. Central-government departments have held different views on the problem. Industry, for example, has traditionally taken a somewhat agnostic line on the whole issue. Different regional offices of such key departments as the Environment adopt different stances towards the cities. Once the argument is widened to include other important agencies, such as local government and the private sector, the scale of the problem begins to emerge.

It has been virtually impossible to achieve a corporate response from interested parties towards the problems of the inner cities. Indeed the evolution of inner-city initiatives from the late 1970s can be interpreted almost as a dialectic between such initiatives as Task Forces and Partnerships (designed to integrate activities) and other innovations, such as Urban Development Corporations that have a more executive role. As the complexities of co-ordination have become ever more apparent, it is the executive type that have come to dominate.

Other issues merit attention, and the question of resources is relevant here. A complex web of inter-relationships funnels resources, primarily from central government and the private sector, to local authorities and voluntary organizations. These relationships can help to explain the development of inner-city policy. For example, despite statements to the contrary from central government, the cities have lost and not gained in resources. Inner-city funds are limited, and in many instances have been sliced from the tops of existing budgets. Throughout the 1980s, in terms of the rate of support grant the cities have lost considerably in both capital and revenue. At the same time, business interests began to play a larger role in the urban debate. By 1988, the Confederation of British Industry was arguing that urban intervention needed sums in excess of £50 billion (Confederation of British Industry, 1988). One interpretation of inner-city policy is that, despite government comments to the contrary, it remains a marginal, underfunded policy area providing little new for the cities. However, while central governments in the 1980s have dismissed criticisms emanating from local government, towards the end of the 1980s, the business sector's determination to highlight the scale of the problem may pose awkward questions for central government.

One of the most important of all explanations for the scale and direction of inner-urban intervention – certainly in the 1980s – is that of central–local relations. However, as this issue has an ideological basis, it is addressed later in this chapter.

Interest Groups

A variety of interest groups has an effect on the direction and intensity of urban intervention. For example, the voluntary sector, locally and nationally through such bodies as the National Council for Voluntary Organizations, has stressed the particular importance of the Urban Programme in funding and promoting community schemes. Local government has also played a real, if diminishing, role. The Association of Metropolitan Authorities has consistently highlighted the scale of, and resources needed to ameliorate, inner-urban problems (Association of Metropolitan Authorities, 1986). Rather surprisingly, so too has the Association of District Councils (Association of District Councils, 1987). Professional groups, such as planners, have used aspects of inner-city funding and legislation to implement a range of projects, for example, Industrial Improvement Areas, and have argued for an enhanced urban dimension (Town and Country Planning Association, 1986).

However, in the 1980s, what might be termed social and professional interest groups have not enjoyed a great deal of success. They have, on the whole, remained in existence, despite what may seem a complete abandonment of collective policy by central government. Moreover, they have achieved relatively minor victories, such as enhanced funding for voluntary groups in many of the

Partnerships. However, bearing in mind the scale of the problem as seen by many of these groups, and the diversity of policy initiatives they have proposed, the impact of these interest groups has been limited.

One obvious exception to this argument is the business community. In the 1980s, the private sector has featured more prominently in a number of urban initiatives. Local chambers of commerce and industry have vetted Partnership programmes; personnel have been seconded to such initiatives as the Task Forces; Business in the Community has emerged as a substantial influence on the urban scene; the CBI has promoted the notion of corporate involvement in the renewal of physical and human capital in cities; and so on. Since 1979, these developments have been welcomed and encouraged by Conservative governments. However, as is discussed more fully in Chapter 8, the increasing involvement of the private sector in urban policy has its advantages and disadvantages. On the one hand, this trend may result in additional skills and resources in the cities, and it has presided over a considerable amount of urban development. On the other hand, however, the more the business community is involved in urban policy, the less likely it is that issues such as deprivation, poverty and equity will figure prominently in the urban debate.

Ideology and Inner-Urban Policy

A number of qualifying statements should be made before examining issues of a more ideological nature. As is mentioned earlier, many observers use both ideological and policy analysis themes to isolate urban intervention: the two approaches are not mutually exclusive. Many ideological assumptions are stated implicitly rather than explicitly. Some ideological explanations do not fit easily into one of the classifications outlined below, and some are not only an explanation for what has happened, but also a justification for what should happen. Despite this, ideological considerations are important and deserve comment. One simple but effective approach is to follow a fourfold division: anti-collectivism, reformism, élitism and radical pespectives.

Anti-Collectivism and the Cities

Modern anti-collectivism had its origins in the thinking of such commentators as Friedman (1962) and Hayek (1979). This approach assumes that the market represents the most efficient mechanism for organizing economic activity. The State should only be peripherally involved in aspects of production and distribution. It may be necessary for the State to regulate markets, to prevent monopolies from occurring and to ensure a stable money supply, but other areas of activity that UK governments since 1946 have been involved in, such as the provision of an extensive welfare system, economic planning, direct State economic operations, and so on, should be reduced to a minimum. In other words, collective activity should be substantially reduced.

These arguments, which had steadily been abandoned throughout this century, re-emerged with the election of Thatcher's 1979 government. This government must have claims to be the most right-wing, anti-collectivist administration to have been elected for many decades. The State was to be 'rolled back', and notions of 'enterprise' and 'freedom' dominated official Conservative attitudes. The post-war political consensus based on demand management of the economy and the provision of an extensive welfare system, was to be reduced. This consensus had been substantially undermined as a result of the 1976 cuts imposed by the International Monetary Fund, but policies promoted by Conservative governments in the late 1970s and early 1980s weakened it still further.

This apparent drift in the late 1970s to anti-collectivism had very real implications for the cities. In theory, it could be assumed that the assimilation of anti-collectivist ideals in central government would result in an immediate and substantial reduction in all forms of State aid for the older conurbations, as at least one independent observer suggests (Rogaly, 1976). The argument is that State regulation and intervention had increased the urban problem: tight planning controls, profligate local councils and widespread regulation had dampened down enterprise and initiative in the older urban cores. These trends would reduce activity in the cities.

Additionally, anti-collectivist logic argued, markets should be allowed to invest when and where they consider most appropriate. If, as seemed evident in the late 1970s and early 1980s, markets wished to decentralize away from cities, then they should be allowed to do so. Attempts to stop them would be fruitless and expensive, and would reduce overall wealth. If factors of production in cities, such as land and labour, dropped as a result of this process, the older conurbations would eventually price themselves back into the market.

In practice, however, not a great deal of this seems relevant to the urban experience of the 1980s. The assumption that costs of factors of production, such as land and labour, would decline as urban economic activity fell, has been generally inaccurate. Land prices fell, but owners did not sell in the hope that costs would rise again. Public-sector land was brought onto the market through land registers (see Chapter 6), but in many cities most vacant land is owned by the private sector. Likewise, although unemployment rose in many cities, labour costs did not, on the whole, fall, because of the national structure of wage bargaining. If some of the economics underlying anti-collectivist approaches towards the cities seemed to lack credibility, the real reason why this ideology was largely irrelevant in structuring urban policy in the 1980s was perhaps political.

A number of factors were at work. During the 1970s, when the Conservatives were in opposition, and after their election in 1979, reformist, even interventionist sentiments remained within their ranks (see, for example, Walker, 1979). Many Conservatives believed that the cities should not be allowed to decline: there was too much private- and public-sector investment in them; Britain, unlike America, did not have the physical resources to allow employment and population to decentralize constantly from the urban cores; and after the riots of 1981, it was hardly politic stance for any government to appear to be under-estimating the scale of the urban problem.

In retrospect, it seems that, for most of the 1980s, anti-collectivist attitudes towards the cities were muted. Instead, what George and Wilding (1976) describe as a form of reluctant collectivism came to dominate government thinking in relation to the cities. Markets were still seen as the most efficient mechanism to organize production, but the State would need to intervene in a number of ways. For example, government subsidy, such as urban development grant, would need to be introduced in order to attract institutional investors to the older conurbations. It would also be important to set up market-dominated development agencies, such as Urban Development Corporations, to create the right physical and financial environment for market investment. Thus, both local and central government would actively encourage investment in the cities through a variety of subsidies and through the provision of a range of such services as education, training, infrastructure, and communications. Market activities would be moderated by positive subsidies designed to encourage urban investment, and planning regulations and controls to discourage decentralization.

At the time of writing, it would, therefore, be reasonable to argue that anti-collectivism fails to provide an appropriate mechanism to explain either the origin or development of urban policy. However, it might well be that, in the 1990s, it becomes rather more central to the debate. This would occur if, as seems distinctly possible, local government is reduced to a marginal supervisory function and the business community takes on a more high-profile role in urban regeneration. State intervention, and certainly local-authority activity, might thus be substantially moderated and collectivist programmes largely abandoned.

Reformism and the Cities

Reformists argue that, although the market may be the best vehicle to organize production and distribution, the State needs to intervene to ease constraints on market activities and to moderate anti-social consequences of private-sector investment. The scale and direction of State intervention is subject to considerable debate. Some argue for relatively-limited State activity and others for a comprehensive programme of intervention to be undertaken by both central and local government. But reformists argue that states do, and perhaps should, react to specific problems, such as the case of the inner cities.

This assumption raises a whole series of subsidiary debates. Different 'states' will identify different problems: a pluralistic electoral system will create contrasting administrations in response to changing electoral circumstances. States may highlight specific 'problems' that may not seem problematic to everyone or which require specific ameliorative action. States may also generate problems. Nevertheless, reformism has important insights for inner-urban policy.

For example, since the late 1960s, it has been clear to both Conservative and Labour governments that something was going wrong in the cities. Unemployment was rising; housing deteriorating; social and physical infrastructure declining; and so on. The problem, therefore, required State intervention. More resource and policy initiatives needed to be directed towards the urban cores. Though the approach adopted by Conservative governments in the 1980s increasingly diverged from that instituted by the Labour government in the late 1970s, these differences, the argument runs, are cosmetic rather than real. Policies instigated in the 1970s have survived into the 1980s, and the approaches adopted by, say, Secretaries of State for the Environment such as Walker in the early 1970s, and Heseltine a decade later, were not in principle very different from those adopted by Labour politicians, such as Peter Shore, in the intervening period.

A consistent theme emerges: an urban problem has been widely identified; an urban policy has emerged; reformist Secretaries of State have introduced, or tried to initiate, programmes designed to enhance economic activity and to reduce poverty in the cities. This reformist approach is certainly one any number of independent observers would like to see implemented. Such organizations as the Trades Union Congress (1988), the Association of Metropolitan Authorities (1986) and the Association of District Councils (1987) have all promoted a reformist urban policy based primarily on increased State planning in and for the cities, and a Keynesian strategy of enhanced public-sector investment.

Many observers might wish to see the introduction of a reformist urban package. More important, however, is the question of whether reformism explains the approaches applied to inner-urban policy. It would be wrong to deny that some central-government ministers have been alarmed at the state of the cities and, as a result, have introduced or tried to instigate policy responses, but whether inner-urban policy can be assessed as a reformist package is open to considerable criticism. This explanation ignores the political benefits that central politicians and central governments anticipated they would receive as a result of intervention. It ignores the wider governmental role in such areas as housing and social security, which have operated to the dis-benefit of many urban residents. Importantly, it avoids the central issue that urban policy has not succeeded in many of its objectives: it has not eroded poverty; housing standards have declined; unemployment is still far too high. One major reason for this is that all governments have not allowed their reformist enthusiasm to undermine larger-scale economic policy. As has been mentioned on numerous occasions in this book, the cities have lost, not gained, resources from central government. Between 1981 and 1987, the seven Partnerships lost more from the rate support grant than they gained from the Urban Programme (Darwin, 1988). This does not seem like reformism in practice.

Élitism and Urban Policy

Élitism is based on the idea that cohesive ruling groups are able to influence or even control the direction of policy. Much of the discussion surrounding the existence or otherwise or élites in the UK has concentrated on local politics

(Dunleavy, 1982). In some authorities there seems to be evidence that a small group of elected and non-elected people is able to structure the political agenda according to their own preferences. This élite may sometimes be held together by economic considerations. For example, in Conservative-controlled authorities there may be close personal ties between politicians and local businessmen; in Labour councils the political agenda may be defined by a small group of local councillors.

However, too much should not be made of the existence or otherwise of local élites: in the 1980s, one dominant trend apparent in British public administration has been the erosion of local power through central regulation and control. If local élites exist, their effect overall is not going to be great in a centralized society.

If élitism has any validity in helping to explain inner-urban policy, it is to the national level that we should turn. Two arguments might have some (admittedly limited) applicability here: first, the development of early urban experiments and perhaps some later urban policies reflect the attitudes and aspirations of a small bureaucratic élite within the Civil Service. A number of commentators have pointed out the essentially pragmatic, *ad hoc*, often unexpected nature of urban experimentation (Edwards and Batley, 1978; McKay and Cox, 1979; Higgins *et al.*, 1983). Potential urban projects emerged initially as a result of the activities of small groups of key civil servants, many of whom had seen similar practice in America and who might gain professionally from the implementation of certain schemes.

The same arguments could be made about later policies, although Secretaries of State like Michael Heseltine were important in promoting innovation. But in general, the idea that the urban experiments resulted from widespread lobbying is unfounded. Throughout the 1960s and 1970s, there was little political demand for the implementation of specific urban projects. The urban dimension's early development may have reflected general disquiet at the state of the cities, and the determination of some ministers to effect an inner-city programme, but the role of a small bureaucratic élite was important in structuring intervention.

A second way in which élitism could help to explain the origin and development of inner-urban policy is through the increasing involvement of the business community in ordering and effecting policy (see Chapter 8). The argument is that a business élite has encouraged the corporate sector to take a greater interest in the older conurbations. Such developments as the Confederation of British Industry's Urban Task Force and Business in the Community have been promoted by larger British companies. These initiatives have been recognized and, it might be argued, have been rewarded by Conservative central governments in the 1980s. As *Action for Cities* (HMSO, 1988a) makes clear, the private sector should act as a prime agency in creating urban regeneration and is to be encouraged to do so through a number of public-sector subsidies.

The business community, particularly the larger corporations, have had an increasing role to play in defining urban intervention. Whether this is élitism in action is open to debate. The argument should not, perhaps, be taken too far: the business community's programmes and aspirations for the cities are not very different from those supported by central government after 1979.

Radical Interpretations of Urban Policy

The last section of this chapter attempts to outline radical explanations of inner-urban intervention and it needs to be stressed that this is an outline – the sheer volume of material emerging from, or critical of, radical explanations of the urban problem is daunting. Those interested in the area should turn to other sources for a fuller discussion (Dear and Scott, 1981; Friend and Metcalf, 1981; Cooke, 1983; Rees and Lambert, 1985; Lawless, 1986).

Most radical explanations have come from neo-Marxist thinking. Central to these is the class-based nature of a capitalist economy founded on the process of accumulation. Accumulation creates profits, ensures re-investment and thwarts competition. However, a series of crises affects the capitalist economy. Some of these crises may occur because of an over-concentration of profitability in larger companies to the detriment of other sectors in the economy; others because of under-consumption; and others because of rising costs in factors of production, such as labour; and so on. Whatever the causes of crises in the capitalist economy, the net result (according to some commentators) tends to be declining profitability, unemployment, diminishing investment and the possibility of a collapse of the capitalist system and its replacement by an alternative socialist society.

There are many variations on this theme. Many observers point to the dynamic nature of change within the economy, rather than its leading inevitably to the end of capitalism. Analyses of capitalism in the late 1980s may find it hard to conclude that the system is about to disappear. As an economic system it has undoubtedly created inequalities and imposed severe costs on some individuals, communities and even countries. It has suffered from periodic crises of profitability and output, but many have gained materially from it and it would appear to be the dominant economic system in the world.

Although radical explanations of the economy are, at the very least, open to some debate and discussion, the approach as a whole provides some valuable perceptions for exploring inner-urban policy. Two concepts are particularly significant here: the role of the State and central–local relations.

The role of the State

Radical observers generally argue that the State has an important role to play in the prevailing economic system. Its exact function, however, is open to debate – between those regarding the State as a mere instrument of capital, and those allowing it a degree of autonomy. But whatever its ultimate power, the State has a twofold function: it must ensure the continued accumulation of profitability in the system while at the same time ensuring the legitimacy of that system to the populace. This tension offers a number of important possibilities to those trying to isolate inner-urban policy.

The entire edifice of inner-urban intervention can be seen as a notable instance of the legitimation/accumulation dilemma (King, 1987). On one hand, such aspects of the urban dimension as, for example, welfare and social spending under the Urban Programme, can be regarded as attempts by the State to legitimize the economic system to the more deprived, and potentially riotous, communities in the older conurbations. On the other much of what has happened in the 1980s in inner-city policy can be assessed as the State intervening to secure continued accumulation. By the late 1980s, much of the inner-city dimension was about providing State subsidies to corporations involved in either development or production or both. Thus the State is attempting to ensure continued accumulation within the system by reducing the costs to the market and boosting profits.

Within this broad legitimation/accumulation concept, the debate about urban policy can be taken much further. A number of commentators are relevant here. Habermas (1976) argues that crises in the capitalist system trigger off a variety of additional problems. Economic crises create fundamental problems, since the State is usually incapable of intervening effectively in the economy – what Habermas terms the rationality crises. In turn, this failure leads to legitimation and motivation crises, in which many of the economically disaffected lose their trust in the prevailing economic system, possibly resorting to active discontent, thus stimulating the State into coercive practices.

This concept has important bearings on the British urban experience. Urban policy can be regarded as a classic rationality problem for governments elected since 1979. The State has been unable to effect a coherent strategy towards the cities. The difficulties of implementing a genuinely-innovative, co-ordinated and integrative programme have been too great. In a similar fashion, the urban disturbances of the 1980s can be seen as instances of legitimation and rationality crises. There are too many people in the cities who have little confidence in prevailing economic and political systems, and they have sought out alternative life-styles and, on occasion, they have reacted violently towards what they consider the dominant inequalities determining their lives. In a further development, O'Connor (1973) argues that the twin pressures of accumulation and legitimation require governments to undertake various kinds of State investment. Some investments are social expenses allocated to functions that are economically important but that do not contribute directly to profits. Other State expenditure – social capital – contributes to profitability since it represents provision of resources for activities that the private sector would otherwise have had to fund.

O'Connor's typology of social capital is complex, in part representing investment into fixed capital. This expenditure is complementary or discretionary. Complementary expenditure is designed to reduce costs to the market, or to provide necessary resources for firms not provided for elsewhere. Saunders (1982) argues that in the urban context there are three aspects to complementary investment: infrastructural provision, direct subsidies and State shareholding. All three have figured in the urban dimension. The Urban Programme, for example, has provided such physical infrastructure as road improvements; subsidies have been forthcoming through, say, the Urban Development Corporations; and the enterprise boards have provided direct equity shareholding. Discretionary investment is provided by the State to encourage equivalent or greater investment by the private sector. Urban development grants and the whole principle of leverage provide a notable example of this kind of investment.

Central–local relations and inner-urban policy

The final insight on the urban problem to arise from the radical debate is perhaps the most revealing to students of inner-city policy. Central–local relations have increasingly come to figure in the urban debate, and full discussions can be found elsewhere (see, for example, Duncan and Goodwin, 1988). At the core of this analysis is the idea that, in the 1980s, central government has dramatically increased its control over local authorities. It has done this for a variety of reasons. Conservative governments have not been impressed by the interventionist strategies adopted by a number of authorities in local economic development, housing policies, subsidies on public transport, and so on. Central government has been opposed to the radicalization of councillors and officers in some urban Labour-controlled authorities.

Local government has been increasingly considered as relatively inefficient by central government ministers: Urban Development Corporations, for example, were regarded as more able to bring a focused and single-minded approach to urban renewal than was possible through local government (Darwin, 1988). Increasing control by central government has also occurred because it allows the government to intervene in what have traditionally been regarded as locally-determined issues. Such policy areas as housing, education and transport have seen dramatically-increasing central control, with local government being steadily relegated to a regulating rather than implementing authority. Increased central control also allows the government to experiment and to evaluate projects at a local level rather than to implement national policies. This localized approach is cheaper, and it ensures damage limitation if the experiment fails. Finally, central governments have intensified their control of local government because governments elected since 1979 have acquired an increasingly autocratic nature – dissent has not been encouraged from any quarter.

Increased central control provides a useful framework in which to explore inner-urban policy of the 1980s. Almost every innovation introduced by Conservative governments after 1979 has involved a loss of locally-determined control. Urban Development Corporations remove development and planning powers from councils; the new City Grant by-passes local government; the Task Forces remain shadowy entities with little local democratic accountability or representation. Similarly, enterprise zones reflect central government's willingness to experiment locally. The original idea may have been substantially watered down, but essentially the concept was about the effect of liberalization on economic activity in defined areas. Inner-urban policy would appear to provide a notable instance of heightened central control.

While this assumption could be perceived as generally accurate, it merits comment and refinement. For example, not all radical commentators have considered local government as representing a genuinely-progressive alternative to that provided by the State (Cockburn, 1977). There are debates about how far local government can act, or would wish to operate, as an autonomous entity (Dearlove, 1979; Saunders, 1981). A substantial number of Conservative-controlled district councils have welcomed such government proposals as those that allow councils to sell public housing. The number of councils that have consistently pursued radical local strategies is very small, especially since the abolition of the metropolitan counties.

It has been argued that, although central government has increased its control, the process is more complex than a power struggle (Rhodes, 1980). There are degrees of negotiation and bargaining between the central and local government, and local government still retains a degree of discretion. These observations apply to inner-urban policy. A few initiatives, notably the Urban Programme, still incorporate local government. Partly as a result of local-government lobbying, some schemes, notably enterprise zones, were substantially refined from how they were originally conceived.

There has been an element of discretion in this. In some instances, projects are permissive – it has been up to councils whether they wanted a specific initiative or not. With the rise in prominence of Urban Development Corporations, this discretion has been less obvious. It would be incorrect, on the other hand, to imagine that all councils have objected to the imposition of an Urban Development Corporation – some of the later corporations were welcomed by the authorities concerned, not least because negotiated agreements with corporations proved possible in some instances.

This development may lead to a conclusion that, although centralization has proved the dominant pattern to central–local relations in the 1980s, a form of local corporatism has also emerged (Cawson, 1985). Local authorities, according to this view, have acted increasingly as managers within their administrations, putting together development packages with other local agencies, such as the business community, trade unions, academic institutions and even Urban Development Corporations. From this point of view, local authorities have not been dominated by central government. Rather central government regulations and controls affecting urban councils wishing to enter new relationships with other local agencies, have given rise to a form of local corporatism: a synthesis of local interests with the common goal of economic development.

Despite these reservations, no examination of inner-urban policy should ignore the question of central control – it is vital to an understanding of the evolving nature of the urban dimension. In the 1970s and early 1980s, conspiratorial evaluations of inner-city intervention appeared inappropriate in the context of a marginal and pragmatic policy field (Bridges, 1975). At the time of writing, it becomes increasingly clear that central government has employed inner-urban policy to extend its own power locally, while at the same time diminishing the role of local government. This message was at the heart of the 1988 statement, *Action for Cities* (HMSO, 1988a) and an implicit, if not explicit, consequence of Urban Development Corporations, city grants, Task Forces, and other initiatives.

Conclusions

The final task remaining is to identify the explanations that most obviously 'fit' the origin and development of British inner-urban policy. Three influences operating on different levels seem to hold important clues. On a basic level, the role of politics and politicians has been influential – in the late 1960s, the Wilson government's determination to do something about the cities, and to counteract Powell's racist speeches, was vital in creating the initial urban dimension. Certain Secretaries of State for the Environment, notably Michael Heseltine, have given the area greater substance and status. More generally, the area has acquired increased political significance in the late 1980s, which will sustain the urban dimension well into the 1990s.

On an intermediate level, questions about central–local relations must feature prominently in any evaluation of inner-city intervention. Whether, in the 1980s, central governments have deliberately sought to enhance urban policy because it allows increasing control of locally-determined issues is debatable, but this appears to be what it has done.

Finally, on a higher level, inner-city policy is a notable instance of the conflicts and tensions between the State's functions, on the one hand, of ensuring accumulation while, on the other, of legitimizing the prevailing economic system. One factor influencing central governments in the late 1970s and throughout the 1980s, was the scale of economic retrenchment in the cities: unemployment and output declined drastically between 1979 and 1983. Increasing profitability and output in the cities required State intervention at a time when the market was disinclined to invest in most of the older conurbations – hence, urban development grants, subsidies through Urban Development Corporations, Industrial Impovement Areas, and so on, were introduced.

At the same time, however, problems of legitimation grew larger, and were most evident in the urban distubances of 1981. This resulted in the perpetuation of the Urban Programme, with its support for social, community and voluntary projects; in efforts to develop community policing; in educational projects designed to link pupils to local jobs; in targeting jobs created as a result of urban regeneration projects on local residents, and so on. The tension between legitimation and accumulation provides an illuminating framework to locate much of what has happened in and to the cities.

11
Inner-City Policy: Towards Reform

Inner-urban policy has failed to reverse the processes that have undermined the social, economic and physical fabric of many major British conurbations over the last twenty years or so. With a programme as divergent as inner-city policy, there will inevitably be some successes: a great deal of physical development has occurred within some Urban Development Corporations and enterprise zones, and many worthwhile individual projects have been funded through the Urban Programme. However, it would be difficult to find observers who believe that the series of *ad hoc* projects implemented since the late 1970s represent a genuinely-coherent attack on the cities. Academics (Stewart, 1987), the Association of Metropolitan Authorities (1986), journalists (Harrison, 1983), independent organizations (Town and Country Planning Association, 1986) churches (Archbishop of Canterbury's Commission on Urban Priority Areas, 1985), the Association of District Councils (1987), business organizations (Confederation of British Industry, 1988) and even Conservative politicians (*Guardian*, 1988) have all, to varying degrees, regarded inner-city policy as inadequate to deal with contemporary urban problems. What sort of policies might be introduced that would create a coherent urban programme?

One way is to explore potential policy developments in a hierarchical framework. In Chapters 9 and 10, the analysis of inner-city policy is structured on three levels of debate. Lower-level issues concern how urban policy is effected; middle-level analysis is concerned with institutional considerations; and highest-level debate concentrates on ideological issues. This threefold division provides a structure within which to discuss issues of policy.

Urban Policy: a New Direction

Urban policy has not been characterized by support for innovative projects. Change has occurred, and some apparently successful American approaches adopted, for example, the principle of leverage. But the vitality and creativity in many local councils, community groups and other non-statutory bodies has not been reflected in inner-city policy. Policy all too often appears to be a bureaucratic blanket smothering innovation in a determined effort to keep to budgets and deadlines, and to respond to the demands of monitoring.

Evaluation is needed, but it should be more effective. It should not be over-concerned with the details of particular projects and schemes: it should address broader issues, notably the degree to which initiatives achieve meaningful objectives in such areas as employment. Creating so many jobs is not enough – the questions needing to be asked, among others, are, jobs for whom, at what net cost, for how long and at what cost to other producers? These considerations are beginning to appear in some analyses, particularly of employment. They need to be widened to include other policy areas, such as housing, deprivation, social services and education. However, even if this were achieved, the fundamental problems affecting inner-urban policy would scarcely have been addressed. The causes of many in adequacies in inner-city policy do not lie in implementation – the programme emanates from a complex inter-play of institutional and ideological forces.

Institutional Change and Urban Policy

Inner-urban policy derives from a range of institutional processes and constraints. If there is to be a sensitive policy towards the cities, unprecedented and, it may be argued, unrealistic change will be needed in its strategy.

Strategy and Inner-Urban Policy

Inner-city policy has been dominated by semi-autonomous agencies and random initiatives to the detriment of longer-term strategy. This process was not helped by the abolition of the metropolitan counties and the establishment of unitary plans that are not going to provide coherent and comprehensive guides to urban development over the next decade or so. A broader strategic vision is vital: cities need to decide whether inner-urban policies should look to such new employment sources as tourism and leisure; the extent to which vacant industrial land could be used for housing or for open space; the relationship between the older urban cores and the surrounding commuting zones; the scale and direction of investment in public transport; new administrative structures that could be used to encourage community-based enterprises; the role of the private sector in devising and implementing urban regeneration; the role of education and training within the new urban economies; and so on.

The list any strategic urban statement needs to address is formidable. Some considerations will have already been incorporated into Urban Programme submissions, corporate planning documents, structure plans, and so on, but these rarely include the full range of pertinent issues over appropriate time-scales.

Strategic and costed urban strategies need to be devised and agreed by local and central government, working in co-operation with other interested parties.

Different cities require different regenerative programmes, but inner-urban policy has tended towards the uniform. Planned decentralization can be used as an example. In many cities it could be argued that decentralization should play a minimal role in alleviating urban problems – sensitive replanning of the older urban cores could proceed without the need to move people and jobs out. Whatever the validity of this – the current orthodoxy – it is interesting to note that the two most substantial studies of London's needs undertaken after 1977 have both suggested that planned decentralization from the capital should be positively encouraged (DoE, Lambeth Inner Area Study, 1977; Buck, Gordon and Young, 1986). This is not the place to debate the relative merits of this position. It is worth pointing out, however, that unless strategic urban statements are forthcoming, the variety in approach that will need to be adopted to cater for the differing needs of the conurbations will tend to be lost within a sea of ad-hocery. *decentralize – reorganised into smaller units.*

A Comprehensive Urban Stategy

On the whole, inner-urban policy has released small amounts of money to stimulate development projects. There have of course been other aspects to the urban dimension. Any Urban Programme submission would reveal a wide range of social, environmental and economic schemes receiving support. But the sums involved are insignificant. And certainly if *Action for Cities* is to be regarded as the government's urban strategy for the late 1980s and into the 1990s, the emphasis is undeniably on aspects of development. While development should play a role in the inner-city programme, it is equally apparent that urban intervention has not assimilated as many other policy arenas as it should have done. Some indication can be given of the range of initiatives that could be introduced or substantially improved.

- *Employment* Job-creation programmes based on increased investment in such areas as social services, education, construction and infrastructural investment; boosting community and co-operative enterprises; job subsidies for the young and long-term unemployed; support for ethnic-minority enterprises; contract compliance; targeting jobs at local residents.
- *Housing* Greater support for co-operatives; more resources for improvement and construction; an increase in lower-density dwellings, some at the top end of the market.
- *Social and community welfare* An improved system of income maintenance and welfare support; additional resources to deal with problems associated with single-parent families and the elderly; more crèches and facilities for the young; more effective take-up campaigns; ethnic-support systems.
- *Education and training* Extended nursery and adult education; grants for

16–18-year-old students in need; a substantial across-the-board boost to educational and training investment, not simply in city technology colleges, and so on.

- *Health* Many diseases and illnesses are more apparent in the cities than elsewhere – more resources are needed, particularly in terms of community health and prevention.
- *Development* Sites need to be made available from public and private sector landowners to cater for demand from existing and new employers.
- *Infrastructure* Sewerage and other systems are collapsing – a ten-year action programme is required, which would also help to boost urban employment.
- *Transport* Investment has declined in recent years; a new approach is needed that, while improving roads, provides current expenditure to support existing provision and capital investment to construct rapid-transit systems.
- *Environment* Ecological areas could be devised; more open space created; cycleways and footpaths installed.
- *Police* Policing needs to be brought to the forefront of the inner-urban debate; recruitment policies should reflect the population of inner-urban areas; positive steps should be taken to eradicate racism within the police; greater police accountability.
- *Decentralization* Too many inner-city residents are trapped, often in public- and private-rented accommodation; they cannot move to the suburbs and smaller towns and cities where better environmental and economic conditions prevail because the only housing being built beyond the cities is in the owner-occupied sector. If the over-concentration of the poor and deprived within the cities is ever to be moderated, new approaches based on the provision of rented accommodation, housing association dwellings, and on the substantial development of new tenures, such as equity sharing, need to be pursued more vigorously.

Attempts to generate such policy objectives as these will inevitably create difficulties. Problems of co-ordination, discussed below, would greatly increase. And it may be naïve to assume that British public administration can deal with problems such as urban decline which clearly cut across traditional departmental briefs. But that should not be allowed to hide other realities: inner-urban problems are wide ranging, complex and inter-related. They will not be solved by an overwhelming emphasis on providing public subsidy for development projects. Conservative administrations may wish to equate success with the degree to which development projects are implemented in the older urban cores. There are other criteria such as equity, by which the urban dimension needs to be assessed.

Inner-Urban Policy: Administration and Co-ordination

Administrative innovation is required locally, regionally, nationally and in relation to development agencies. Local government is the natural agency through which much inner-city innovation should be channelled: councils know about the problems of their administrations and they are in an ideal position to integrate contributions from the private sector and from the community. They are also the elected representatives of their local communities. However, it would be wrong to say that local authorities have always been efficient.

Efficiency is an important issue in that a great deal of inner-urban investment in such areas as housing and economic development will have to be levered from the market. It is therefore important that councils are in a position to stimulate market investment. Recent evidence from such cities as Glasgow and Sheffield has shown that councils are becoming much more aware of the attitudes and institutional innovations they need to adopt if they are to acquire substantial private-sector investment.

In addition, two other, more formal, developments should be encouraged. First, as is discussed above, comprehensive urban strategies need to be adopted by local authorities, in which specific sections are dedicated to collaborative programmes agreed with the market and community groups. Community groups have been neglected in inner-city policy. Yet, if the political alienation in the older urban cores is to be moderated, new forms of locally-based representative government need to be installed. Second, local government should be given every encouragement to create separate companies to achieve specific ends. Enterprise boards have shown how successful this approach can be.

Much of the debate about how inner-urban policy should develop has concentrated on either the local level or initiatives devised by central government. There is another important strand of administration: the regional dimension. Elected regional government does not exist in Britain, however. At various times, all political parties (with the exception of the Conservatives) have supported the idea: the counties would be abolished and about eight regional authorities created in England, with two, more powerful, bodies in Wales and Scotland. This would counteract the over-concentration of power in the South East and give greater independence and autonomy to the regions.

This would help the cities in a number of ways. First, it would assist in the decentralization of administrative, financial and media jobs from London and the South East. It is not surprising that the provincial capitals of Edinburgh and Cardiff performed better than virtually any other cities in replacing lost manufacturing jobs with new service-sector jobs in the 1970s (Hall, 1987). Second, the installation of regional goverment would lead to a more rational system of resource planning. The regions would need to produce development strategies governing such issues as major infrastructural provision, transport, technological innovations, education and training, and so on. These broad policy statements would have to be approved by central government and they would inform the planning of the cities. This would not create more planning – structure planning in the counties would disappear .

Third, central government would indicate the resources to be spent on each region – an important consideration for cities in that many are located in regions that have fared badly in many aspects of national expenditure. Fourth, the regions would have powers to devise their own development strategies. In some respects, for example, inward investment and infrastructural investment, the regions are in a better position to sustain appropriate policy developments than are the cities.

The most significant administrative changes will be required not at a local or regional level but in central government. It appears virtually impossible for one minister, whether Secretary of State for the Environment or for Trade and Industry, to produce a genuinely-integrated programme of support for cities from the large number of departments that might be involved in some way or other in urban regeneration. A separate department is needed with specific responsibilities for the older industrialized regions and cities of England. All departments in turn should indicate their resources allocated to, and policies designed to enhance the fortunes of, the older regions and cities. A principled stance should be taken to reallocate a proportion of departmental resources/manpower to the regions within a specified time period. Bland assertions from the centre that the fortunes of the older regions and cities are taken into account will not do. A commitment to identifiable change over a specified period of time is required.

Chapter 6 shows that the early evolution of the London Docklands Development Corporation was surrounded with controversy. However, experience in Scotland with the Scottish Development Agency, suggests that there is a role in England for a development corporation. Ideally, an English Development Agency would not be solely concerned with development in relatively small parts of the major cities. Instead, it should be mobile and charged with a wide range of functions, including not only economic and physical development but also employment creation, environmental improvements, equity investment, training and possibly even housing. It would only operate in conjunction with appropriate urban and regional councils. It could have a particular role to play, for example, when major plant closures occurred or when large infrastructural projects required supportive investment.

Ideology, Politics and Urban Policy

At the time of writing, it seems certain that many of the policies outlined in this chapter are not going to be implemented in the short to medium term. The dominant political forces in the UK suggest that the cities and declining regions generally are unlikely to benefit from any central largesse. Conservative governments may govern Britain for many years, with a Labour Party in opposition unable to gain seats in the south of England, and with a divided centre-left opposition. The most favourable prospect for the older cities and regions, for the elderly and the poor, and for what remains of the Welfare State, is not so much a change in government but change in Cabinet personnel.

The implication of permanent Conservative government seems to have instilled into many observers the belief that the policies the government has implemented since 1979 were in some way inevitable – that after years of steady but slow decline hard remedies were needed. In some respects, the ideas promoted by governments after 1979 have proved popular and, it could be argued, necessary: for example, there was little sympathy within the country as a whole for the type of corporatist State the Labour government of the 1970s tried to pursue. However, assumptions that hard policies have to be implemented, that there are no alternatives and that policies were in some way objectively and intrinsically correct are untenable. There were and are alternative policies. For the cities, strategies pursued by Conservative governments since 1979 have proved unfortunate.

This has been particularly apparent in the macro-scale economic strategies applied since 1979. In the early 1980s (and there are indications that similar events may occur at the end of the decade) an overvalued pound and high interest rates were extremely harmful to manufacturing industry. In a few years in many conurbations, one-third or more of output and employment was lost. By 1983 the UK had become an importer of manufactured goods for the first time in centuries. There was an argument that British industry needed a shake up: overmanning was evident in some sectors and productivity was often poor compared to other countries. Even if that was the case, the policies effected in the early 1980s were too rigorous. They destroyed whole sections of industry. By the end of the decade, what was left of the industrial base was performing well, catering for a credit-inspired upsurgence, but demand could not be met from a drastically reduced domestic manufacturing sector. Thus a substantial rise of about 15 per cent per annum in imports occurred towards the end of the 1980s, at a time when exports were increasing at about 2 per cent per annum. By 1987, manufacturing output was below totals achieved in the early 1970s.

Different policies could have been implemented. Interest rates were far too high, partly because VAT was increased so much it inevitably led to inflation, the eradication of which was the government's prime objective. Too little was done to support industry – right-wing governments in Europe and the Far East are more supportive of industry's needs: direct financial assistance is available for specific sectors; training is enhanced; research fostered; and government and industry work together to boost output and productivity. Manufacturing jobs are still lost in most industrialized nations, but at more manageable rates and with greater support than is available in Britain.

As part of overall policy, public-sector intervention and resources were also reduced. The public sector was increasingly regarded as both intrinsically ineffective and economically intrusive, in that it crowded out the private sector (Bacon and Eltis, 1976). 'Enterprise' and 'initiative' were key words of the 1980s, and market production came to dominate debate.

As any observer of the workings of some local councils in the 1970s would accept, there was more than some truth in this. Certain public-sector organizations and local authorities were inefficient. There was a need to improve productivity throughout the country in the public and private sectors. However, Conservative governments' approaches solved few of these issues. Productivity rose, although by 1985 output per head was still less than half of that being achieved in America and Japan. Manufacturing output did not increase; the rise in the numbers of self-employed, often seen as an encouraging sign, but frequently reflected job losses in established enterprises; and public-sector spending did not fall substantially. It did, however, change in composition. Such areas as defence, law and social security increased. Other spending budgets, especially housing, education, and local-government finance, fell, sometimes dramatically. In real terms, the rate-support grant from central government to the Partnership and Urban Programme authorities fell by a quarter between 1981–2 and 1985–6. In other words, central-government spending rose, and locally-articulated spending budgets fell.

The effect of this on the cities was profound: local-government spending dropped sharply, with direct implications for such areas as social services and education. Housing budgets were drastically reduced. Between 1979 and 1986, there was a 70 per cent cut in real terms in central government's capital and revenue resources for housing. Not surprisingly, by the mid-1980s the total sum required for improvement, rehabilitation and repair of urban housing stood at around £20 billion (Association of Metropolitan Authorities, 1986). Training was equally reduced: by the end of the 1980s, Britain had 40,000 apprentices, Germany 700,000. Resources for research, higher education, regional development and technical innovation declined in real terms.

This helps to put the inner-city budget into perspective. According to the Association of Metropolitan Authorities (1986), the Urban Programme amounted to 0.0002 per cent of central government's planned expenditure for 1986–7. Resources of this limited nature are not going to counteract substantially the impact of budget reductions elsewhere. It could have been different: a programme for the cities might have been created based on intervention in and collaboration with the market and direct State expenditure.

Private-sector investment needs to be incorporated into urban regeneration: the resources available to the private sector will probably dwarf any that might emerge from the State, even with a change of government. There is, however, another side to the public–private sector equation. As the private sector itself admits, public-sector assistance, especially local-government support, will be necessary in such areas as social services, education, transport provision and – as some market-orientated associations believe – housing (Association of British Chambers of Commerce, 1987; Confederation of British Industry, 1988). This notion of partnership between public and private sectors seems likely to be a central feature of urban regeneration strategies through the 1990s. While it is to be welcomed, it raises at least two issues. First, the private sector is not interested in the public good for its own sake – its motivation is profit. Profitability of individual companies should not be seen to equate with urban regeneration. It is a factor, but many communities and households gain little from corporate profitability. Second, there is an element of hypocrisy in this. The main reason so many cities have endured severe economic retrenchment is because of the corporate restructuring tendencies outlined in Chapter 2. Much of this could have been avoided if a far more effective mergers and takeovers policy had been initially installed (Town and Country Planning Association, 1987). Too many mergers and takeovers make little economic sense, and they destroy long-standing industrial communities. They should only be allowed if a clear economic gain can be identified. Although this idea is politically unrealistic at the end of the 1980s, communities enduring substantial job loss should be compensated by the company concerned. Economic, social and environmental audits should accompany all major plant closures, contractions and even expansions.

The private sector may prove an important agent in urban development in the 1990s, but the real driving force in devising and implementing comprehensive urban strategies must remain with the State. If an effective urban policy is ever to be introduced, central government has to provide additional across-the-board resources. The myraid of projects emanating from central government after 1979 have tended to deflect attention away from the important fact that urban policy requires more State investment. Resources are needed to effect the policies outlined above, to fund local government in a satisfactory manner and to provide the market with the development incentives it will often seek out before embarking on inner-city schemes.

There are no short-cuts here. The 1988 *Action for Cities* statement stressed that new resources for inner-urban investment would not be forthcoming. This is an unfortunate position – the cities are in desperate need of additional funding. In this context the formulation of more *ad hoc* projects, the standard central-government response, will not suffice. A new city technology college or a City Grant may make favourable political ripples, but it will do nothing to alleviate the real problems of the cities. By deflecting attention away from core problems, and by abstracting resources from main budgets, it makes things worse. In this instance there is no alternative: more money is needed.

The standard response to this request is that resources are not available; that the public sector cannot fund any comprehensive urban strategy. As events towards the end of the 1980s show, this argument does not bear scrutiny. Tens of billions of pounds have been spent on keeping an unnecessarily-large number of people unemployed; North Sea oil reserves have done nothing to sustain longer-term growth in the country; funds from privatization programmes have been spent on tax relief – the top 1 per cent of taxpayers receiving tax cuts equal to almost £5 billion between 1979 and 1988; Britain's overseas assets amounted to almost £90 billion by the end of 1987, this at a time when the public-sector borrowing requirement is at an historically low level. Whatever other reasons are given by central government the notion that the country cannot afford a more intensive inner-city dimension is untenable.

Central government can afford more in the way of an inner-urban dimension, but it has been unable and unwilling to initiate it. Implementing an inner-urban programme will never be easy: the complex problems in the major conurbations will necessitate the introduction of an integrated series of policies from a substantial number of organizations. However, central government has never begun to address this issue — its concern has been to create politically-attractive projects, none of which cost a great deal of money and most of which are funded from existing budgets. Questions of policy, of strategy and of equity have disappeared from central government's agenda.

But there are long-term consequences attendant upon a policy that provides so little for so many cities and their poorer inhabitants. High levels in unemployment will remain; infrastructural constraints will become more apparent; and many living in the cities will continue to feel politically and materially alienated. The policies Conservative government pursued after 1979 were inevitably going to increase inequality in Britain. Diminishing State intervention, a restructuring in welfare benefits, market liberalization and fiscal policies designed to benefit the wealthy have led in the 1980s to widening inequalities between a substantial majority who have done well and a disadvantaged minority concentrated in the larger conurbations.

It is morally wrong, economically unwise and politically shortsighted to perpetuate such inequality. Unfortunately, it will probably require more urban disturbances and more middle-class dissatisfaction with prevailing urban services for much to happen. The notion that Conservative governments in the late 1980s are going to take a broader statesmanlike view of the cities and their needs seems unlikely. Change will only occur in response to pressing political dictates. In at least the case of the cities, it might with confidence be predicted that these will not be long in coming.

References

Abel-Smith, B. and Townsend, P. (1965) *The Poor and the Poorest*, Bell & Hyman, London.

Adcock, B. (1984) 'Regenerating Merseyside docks', *Town Planning Review*, Vol. 55, pp. 265–89.

Aitken, P. and Sparks, L. (1986) 'The Scottish Development Agency: a case for co-ordination?', *Regional Studies*, Vol. 20, pp. 476–9.

Aldridge, M. (1979) *The British New Towns – A Programme without a Policy*, Routledge & Kegan Paul, London.

Aldridge, M. and Brotherton, C.J. (1987) 'Being a Programme Authority – is it worth-while?', *Journal of Social Policy*, Vol. 16, pp. 349–69.

Ambrose, P. (1986) *Whatever Happened to Planning?*, Methuen, London.

Anderson, J. (1980) 'Back to the 19th Century', *New Statesman*, Vol. 100, pp. 42–3.

Archbishop of Canterbury's Commission on Urban Priority Areas (1985) *Faith in the City*, Church House Publishing, London.

Association of British Chambers of Commerce (1987) *Inner Cities: Still on: the Agenda*, ABCC, London.

Association of District Councils (1987) *A Blueprint for Urban Areas*.

Association of Metropolitan Authorities (1986) *Programme for Partnership: an Urban Policy Statement*, London.

Aston University (1985) *Four Year Review of the Birmingham Inner City Partnership*. Inner Cities Research Programme 12. Public Sector Management Research Unit for the Department of the Environment, London.

Aston University (1988) *An Evaluation of the Urban Development Grant Programme*. Inner Cities Research Programme, Department of the Environment, HMSO, London.

Audit Commission (1984) *The Impact on Local Authorities' Economy, Efficiency and Effectiveness of the Block Grant Distribution System*, HMSO, London.

Bacon, R. and Eltis, W. (1976) *Britain's Economic Problems: Too Few Producers*, Macmillan, London.

Bains, M.A. (1972) *The New Local Authorities; Management and Structure* (The Bains Report), HMSO, London.

Banfield, E. (1970) *The Unheavenly City*, Little, Brown & Co., Boston, Mass.

Banham, R., Barker, P., Hall, P. and Price, C. (1974) 'Non-plan: an experiment in freedom', in A. Blowers, C. Hamnett and P. Sarre (eds.) *The Future of Cities*, Hutchinson, London, pp. 244–52.

Barnett, C. (1986) *The Audit of War: The Illusion and Reality of Britain as a Great Nation*, Macmillan, London.

Barrett, S. and Fudge, C. (1981) 'Examining policy-action relationship', in S. Barrett and C. Fudge (eds.) *Policy and Action*, Methuen, London, pp. 3–34.

Barrett, S. and Hill, M. (1984) 'Policy, bargaining and structure in implementation theory: towards an integrated perspective', *Policy and Politics*, Vol. 12, pp. 219–40.

Batchelor, C. (1987) 'Co-ops: on a high', *The Financial Times*, 14 July.

Beecham, J. (1978) 'Problems and opportunities of partnership', in T.A. Broadbent (ed.) *Inner Area Partnerships and Programmes: The First Year's Experience*, Centre for Environmental Studies, Policy Series 8, London, pp. 65–70.

Begg, H. and McDowell, S. (1987) 'The effect of regional investment incentives on company decisions', *Regional Studies*, Vol. 21, no. 5, pp. 459–70.

Begg, I., Moore, B. and Rhodes, J. (1986) 'Economic and social change in urban Britain and the inner cities', in V.A. Hausner (ed.) *Critical Issues in Urban Economic Development*, Vol. I, Clarendon Press, Oxford, pp. 10–39.

Benington, J. (1975) *Local Government becomes Big Business – Corporate, Management and the Politics of Urban Problems, Coventry CDP, Final Report Part II*, Coventry CDP, Coventry.

Benington, J. (1985) 'Economic development: local economic initiatives', *Local Government Policy Making*, Vol. 12, no. 2, pp. 3–8.

Benington, J. (1986) 'Local economic strategies: paradigms for a planned economy?', *Local Economy*, no. 1, pp. 7–25.

Beyon, J. and Solomos, J. (1988) 'The simmering cities; urban unrest during the Thatcher years', *Parliamentary Affairs*, Vol. 41, no. 3, pp. 402–22.

Birmingham Council (1986) *Economic Strategy, 1986 Review*, Birmingham.

Birmingham City Council (1987a) *Birmingham Inner City Partnership, Annual Review, 1987*, Birmingham.

Birmingham City Council (1987b) *Birmingham Inner City Partnership, Programme 1987–90*, Birmingham.

Birmingham City Council (1987c) *The Birmingham Training Strategy, Initial Statement*, Birmingham.

Boaden, N. (1982) 'Urban Development Corporations – threat or challenge?', *Local Government Studies*, Vol. 8, no. 4, pp. 8–13.

Boddy, M. and Lovering, J. (1986) 'High technology industry in the Bristol sub-region: the aerospace/defence nexus', *Regional Studies*, Vol. 20, pp. 217–31.

Booth, S., Pitt, D. and Money, W. (1982) *Ambiguity in Action? An Organisational Study of Urban Renewal in Glasgow, Discussion Paper No. 5*, Centre for Urban and Regional Research, University of Glasgow.

Botham, R. (1984) 'Employment subsidies: a new direction for local government economic initiatives', *Regional Studies*, Vol. 18, no. 1, pp. 81–8.

Boyle, R. (1985) ' "Leveraging" Urban Development: a comparison of urban policy directions and programme impact in the United States and Britain', *Policy and Politics*, Vol. 13, pp. 175–210.

Boyle, R. (1988) 'Private sector urban regeneration: the Scottish experience', in M. Parkinson, B. Foley and D. Judd (eds.) *Regenerating the Cities, the UK Crisis and the US Experience*, Manchester University Press, pp. 74–93.

Bradshaw, J., Taylor-Gooby, P. and Lees, R. (1976) *The Batley Welfare Benefits Project, Papers in Community Studies 5*, Department of Social Administration and Social Work, University of York.

Bridges, L. (1975) ' "The Ministry of Internal Security": British urban social policy 1968–74', *Race and Class*, Vol. XVI, pp. 375–86.

Bridges, L. (1981–2) 'Keeping the lid on: British urban social policy 1975–81', *Race and Class*, Vol. XXIII, pp. 161–85.
Brindle, D. (1988) 'Gould claims benefit changes add up to "inner city robbery" ', *Guardian*, 9 April.
British Business (1981) *Statistics*, Vol. 7, no. 1, p. 39.
British Business (1987) *Statistics*, Vol. 27, no. 3, p. 29.
British Business (1988) *Statistics*, Vol. 28, no. 8, p. 38.
Broadwater Farm Inquiry (1986) *Broadwater Farm Inquiry Report*, HMSO, London.
Bruegel, I. (1987) 'Local economic strategies and service employment', *Local Economy*,. Vol. 1, no. 4, pp. 35–46.
Buchanan, G. (1986) 'Local economic development by community business', *Local Economy*, no. 2, pp. 17–28.
Buck, N., Gordon, I. and Young, K. (1986) *The London Employment Problem*, Clarendon Press, Oxford.
Buck, N., Gordon, I. and Young, K. (1987) 'London employment problems and prospects', in V.A. Hausner (ed.) *Urban Economic Change, Five City Studies*, Clarendon Press, Oxford, pp. 99–131.
Burns, W.C. (1980) *Review of Local Authority Assistance to Industry and Commerce, Report of the Joint Group of Officials of Local Authority Associations and Government Departments (under the Chairmanship of W. Burns)*, Department of the Environment, London.
Business in the Community (1987) *Business and the Inner Cities*, London.
Business in the Community (1988) *The Future of Enterprise Agencies*, London.
Butler, P. and Williams, R.H. (1981) 'Inner city partnerships and established policies: the Newcastle/Gateshead experience', *Policy and Politics*, Vol. 9, pp. 125–36.
Callaghan, J. (1968a) *Hansard*, 769, Col. 40, 22 July.
Callaghan, J. (1968b) *Hansard*, 774, Col. 1107, 2 December.
Camina, M.M. (1974) *Local Authorities and the Attraction of Industry*, Pergamon, Oxford.
Campbell, M., Hardy, M., Healey, N., Stead, R. and Sutherland, J. (1987) 'The economics of local jobs plans', *Local Economy*, Vol. 2, no. 2, pp. 81–91.
Castle, B. (1980) *The Castle Diaries*, Weidenfeld & Nicolson, London.
Catalano, A. (1983) *A Review of UK Enterprise Zones*, Centre for Environmental Studies, London, Paper 17.
Cawson, A. (1985) 'Corporatism and local politics', in W. Grant (ed.) *The Political Economy of Corporatism*, Macmillan, London.
Central Statistical Office (1981) *Social Trends 11*, HMSO, London.
Central Statistical Office (1987) *Social Trends 17*, HMSO, London.
Centre for Employment Initiatives (1985) *The Impact of Local Enterprise Agencies in Great Britain*, London.
Centre for Local Economic Strategies (1986a) *Enterprise Boards, A Review of UK Experience*, Manchester.
Centre for Local Economic Strategies (1986b) *Workers' Co-operatives in the UK, Briefing No. 2*, Manchester.
Centre for Local Economic Strategies (1987a) *Economic Sense: Local Jobs Plans*, Manchester.
Centre for Local Economic Strategies (1987b) *Enterprise Boards*, Manchester.
Centre for Local Economic Strategies (1987c) *Local Authorities and Worker Co-operatives*, Manchester.
Centre for Local Economic Strategies (1987d) *More than Just a Poll Tax*, Manchester.
Chalkley, B. (1979) 'Redevelopment and the small firm: the making of a myth', *The Planner*, Vol. 65, no. 4, pp. 148–51.

Champion, T. (1987) 'Momentous revival in London's population', *Town and Country Planning*, Vol. 56, no. 3, pp. 80–2.

Champion, T., Coombes, M. and Openshaw, S. (1983) 'A new definition of cities', *Town and Country Planning*, Vol. 52, no. 11, pp. 305–7.

Chandler, J.A. and Lawless, P. (1985) *Local Authorities and the Creation of Employment*, Gower, Aldershot.

Cheshire County Council (1987) *The Development of Cheshire County Council's Unemployment Strategy*, Chester.

Clarke, A. and Cochrane, A. (1987) 'Investing in the private sector', in A. Cochrane (ed.) *Developing Local Economic Strategies*, Open University Press, Milton Keynes.

Cochrane, A. (1983) 'Local economic policies: trying to drain an ocean with a teaspoon', in J. Anderson, S. Duncan and R. Hudson (eds.) *Redundant Spaces in Cities and Regions? Studies in Industrial Decline and Social Change*, Academic Press, London, pp. 285–311.

Cochrane, A. (1986) 'Local employment initiatives: towards a new municipal socialism?' in P. Lawless and C. Raban (eds.) *The Contemporary British City*, Paul Chapman Publishing, London, pp. 144–62.

Cochrane, A. (1988) 'In and against the market? The development of Socialist economic strategies in Britain, 1981–86', *Policy and Politics*, Vol. 16, no. 3, pp. 159–68.

Cockburn, C. (1977) *The Local State: Management of Cities and People*, Pluto Press, London.

Colenutt, B. (1987) 'UDCs, a challenge to local government', *Local Work*, no. 6, pp. 10–11.

Confederation of British Industry (1988) *Initiatives Beyond Charity, Report of the CBI Task Force on Business and Urban Regeneration*, London.

Cooke, P. (1983) *Theories of Planning and Spatial Development*, Hutchinson, London.

Corby District Council (1981) *A guide to the Corby Enterprise Zone*, Corby.

Cornforth, C. and Lewis, J. (1985) *The Role and Impact of Local Co-operative Support Organisations, Co-operatives Research Unit, Monograph no. 7*, Open University Press, Milton Keynes.

Coventry City Council (1987) *Economic Development and Employment Initiatives in Coventry*, Department of Economic Development and Planning, Coventry.

Coventry Community Development Project (1975) *Coventry and Hillfields: Prosperity and the Persistence of Inequality*, Vol. I, Coventry.

Cowie, H., Harlow, C. and Emerson, R. (1984) *Rebuilding the Infrastructure, The Needs of English Cities and Towns*, Policy Studies Institute, London, no. 633.

Crawford, C. (1986) 'The legal aspects of Section 137 – recent developments', *Local Government Policy Making*, Vol. 12, pp. 70–84.

Crawford, P., Fothergill, S. and Monk, S. (1985) *The Effect of Business Rates on the Location of Employment*, Department of Land Economy, University of Cambridge.

Cullingworth, B. (1974) 'Social problems in cities', in J. Brand and M. Cox (eds.) *The Urban Crisis, Social Problems and Planning*, Royal Town Planning Institute, London, pp. 2–18.

Cullingworth, J.B. (1985) *Town and Country Planning*, Allen & Unwin, London, ninth edn.

Curtice, J. (1987) 'Must Labour lose?', *New Society*, Vol. 80, no. 1277, pp. 17–19.

Danson, M.W., Lever, W.F. and Malcolm, J.F. (1980) 'The inner city employment problem in Great Britain, 1952–76, a shift share approach', *Urban Studies*, Vol. 17, pp. 193–210.

Darwin, J. (1988) 'The need for a new strategy for the inner city', *Local Government Policy Making*, Vol. 15, no. 2, pp. 13–25.

Davies, H. (1987) 'Capital spending', in M. Parkinson (ed.) *Reshaping Local Goverment*, Policy Journals, Hermitage, Berks., pp. 25–37.

Dear, M. and Scott, A.J. (1981) *Urbanisation and Urban Planning in Capitalist Society*, Methuen, Andover.

Dearlove, J. (1979) *The Reorganization of British Local Government*, Cambridge University Press.

Deloitte Haskins and Sells (1984) *Local Enterprise Agencies*, London.

Department of Employment (1988) *Employment Gazette*, Vol. 96, no. 3, HMSO, London.

Department of the Environment (1975) *Study of the Inner Areas of Conurbations*, HMSO, London.

Department of the Environment (1978) *Inner Urban Areas Act, 1978*, HMSO, London, circular 68/78.

Department of the Environment (1980) *The Proposed LDDC*, HMSO, London, memorandum.

Department of the Environment (1983) *Garden Festivals*, HMSO, London, advice note.

Department of the Environment (1984) *Industrial Development*, HMSO, London, circular 16/84.

Department of the Environment (1985a) *Development and Employment*, HMSO, London, circular 14/85.

Department of the Environment (1985b) *Enterprise Zone*, HMSO, London, information 1983/4.

Department of the Environment (1985c) *The Urban Programme 1985*, HMSO, London.

Department of the Environment (1986) *Assessment of the Employment Effects of Economic Development Projects Funded by the Urban Programme*, HMSO, London.

Department of the Environment (1987a) *Enterprise Zone*, HMSO, London, information 1985/6.

Department of the Environment (1987b) *Managing Workspaces*, HMSO, London.

Department of the Environment (1988a) *Creating Development Trusts*, HMSO, London.

Department of the Environment (1988b) *Developing Businesses*, HMSO, London.

Department of the Environment (1988c) *Local Authorities' Interests in Companies, A Consultation paper*, HMSO, London.

Department of the Environment (1989) *DoE Inner City Programmes 1987–88: A Report on Achievements and Developments*, HMSO, London.

Department of the Environment and Department of Employment (1987) *Action for Cities*, Central Office of Information, London.

Department of the Environment, Birmingham Inner Area Study (1977) *Unequal City*, HMSO, London.

Department of the Environment, Lambeth Inner Area Study (1977) *Inner London, Policies for Dispersal and Balance*, HMSO, London.

Department of the Environment, Liverpool Inner Area Study (1977) *Change or Decay*, HMSO, London.

Department of Trade and Industry (1983) *Regional Industrial Development*, HMSO, London, Cmnd. 9111.

Department of Trade and Industry (1988) *DTI – The Department for Enterprise*, HMSO, London, Cmnd. 278.

Docklands Consultative Committee (1985) *Four Year Review of the LDDC*, Greater London Council.

Docklands Consultative Committee (1988) *Urban Development Corporations, Six Years in London's Docklands*, London.

Docklands Forum (1987) *Housing in Docklands*, London.

Duffy, H. (1988) 'Automatic development grants to go', *The Financial Times*, 28 January.

Duncan, S. and Goodwin, M. (1988) *The Local State and Uneven Development*, Polity Press, Cambridge.

Dunleavy, P. (1982) *Political Issues and Urban Policy Making, Unit 18 D 202, Urban Change and Conflict*, Open University Press, Milton Keynes.

Edwards, J. and Batley, R. (1978) *The Politics of Positive Discrimination*, Tavistock, London.

Eisenschitz, A. and North, D. (1986) 'The London Industrial Strategy', *International Journal of Urban and Regional Research*, Vol. 10, no. 3, pp. 419–40.

Estates Times (1987) 'Welcome to resurrection city', 27 July.

Etherington, D. (1987) 'Local economic strategies and area based initiatives – another view of improvement areas', *Local Economy*, Vol. 2, no. 1, pp. 31–7.

Fagg, J.J. (1980) 'A re-examination of the incubator thesis; a case study of Greater Leicester', *Urban Studies*, Vol. 17, pp. 35–44.

Financial Times, The (1986) 1 October, 'London Docklands'.

Financial Times, The (1987) 'Business in the community, survey', 17 July.

Fothergill, S. and Gudgin, G. (1982) *Unequal Growth, Urban and Regional Employment Change in the UK*, Heinemann Educational, London.

Fothergill, S., Kitson, M. and Monk, S. (1985) *Urban Industrial Change, Inner Cities Research Programme, no. 11*, Department of the Environment, Department of Trade and Industry, HMSO, London.

Fothergill, S., Monk, S. and Perry, M. (1987) *Property and Industrial Development*, Hutchinson, London.

Friedman, M. (1962) *Capitalism and Freedom*, University of Chicago Press.

Friend, A. and Metcalf, A. (1981) *Slump City, the Politics of Mass Unemployment*, Pluto Press, London.

George, V. and Wilding, P. (1976) *Ideology and Social Welfare*, Routledge & Kegan Paul, Henley on Thames, Oxon.

George, V. and Wilding, P. (1984) *The Impact of Social Policy*, Routledge & Kegan Paul, London.

Gilroy, P. (1981–2) 'You can't fool the youths . . . race and class formation in the 1980s', *Race and Class*, Vol. XXIII, pp. 207–22.

Greater London Council (1983) *Small Firms and the London Industrial Strategy*, London.

Green, G. (1978) 'Birmingham Partnership: participation excluded?' *The Planner*, Vol. 64, p. 76.

Gregory, D. and Martin, S. (1988) 'Issues in the evaluation of inner city programmes', *Local Economy*, Vol. 2, no. 4, pp. 237–49.

Guardian (1988) 'A strategy for north and south', editorial, 30 April.

Gulliver, S. (1984) 'The area projects of the Scottish Development Agency', *Town Planning Review*, Vol. 55, pp. 332–4.

Habermas, J. (1976) *Legitimation Crisis*, Heinemann, London.

Hall, P. (1977) 'Green fields and grey areas', *Proceedings Royal Town Planning Institute's Annual Conference*, London, pp. 1–12.

Hall, P. (1978) 'Spending priorities in the inner city', *New Society*, Vol. 46, pp. 698–9.

Hall, P. (1982a) 'Enterprise zones: a justification', *International Journal of Urban and Regional Research*, Vol. 6, pp. 416–21.

Hall, P. (1982b) 'Enterprise zones: British origins, American adaptions', *Built Environment*, Vol. 7, pp. 5–12.

Hall, P. (1987) 'The anatomy of job creation: nations, regions and cities in the 1960s and 1970s', *Regional Studies*, Vol. 21, pp. 95–106.

Hall, P., Breheny, M., McQuaid, R. and Hart, D. (1987) *Western Sunrise: the Genesis and Growth of Britain's Major High Tech Corridor*, Allen & Unwin, London.

Hall, P. and Laurence, S. (1981) 'Deprivation in the inner city', in P. Hall (ed.) *The Inner City in Context*, Social Science Research Council/Heinemann, London, pp. 52–63.
Halsey, A. (1972) *Educational Priority. Report of Research Project Sponsored by DES/ SSRC. Vol. 1, Educational Priority. EPA. Problems and Policies*, HMSO, London.
Hambleton, R. (1981) 'Implementing inner city policy', *Policy and Politics*, Vol. 9, pp. 51–71.
Hambleton, R. (1986) *Rethinking Policy Planning*, School for Advanced Urban Studies, University of Bristol.
Hamilton Fazey, I. (1987) *The Pathfinder, the Origins of the Enterprise Agency in Britain*, Financial Training, London.
Hammersmith and Fulham London Boroughs (1987) *Finance and Advice for Small Businesses*, London.
Harrison, P. (1983) *Inside the Inner City: Life Under the Cutting Edge*, Penguin Books, Harmondsworth.
Hart, D.A. (1984) *Attracting Private Investment to the Inner City: the Hackney Demonstration Project*, Joint Centre for Land Development Studies, Whiteknights, Reading.
Haughton, G., Peck, J. and Steward, A. (1987) 'Local jobs and local houses for local workers: a critical analysis of spatial employment targeting', *Local Economy*, Vol. 2, no. 3, pp. 201–7.
Hausner, V.A. (ed.) (1986) *Critical Issues in Urban Economic Development*, Vol. I, Clarendon Press, Oxford.
Hausner, V.A. (ed.) (1987) *Critical Issues in Urban Economic Development*, Vol. II, Clarendon Press, Oxford.
Hausner, V. and Robson, B. (1985) *Changing Cities*, Economic and Social Research Council, London.
Hayek, F. (1979) *The Road to Serfdom*, Routledge & Kegan Paul, Henley on Thames, Oxon.
Hayton, K. (1985) 'Supporting community business', *Local Government Policy Making*, Vol. 12, no. 2, pp. 15–20.
Heclo, H. and Wildavsky, A. (1981) *The Private Government of Public Money*, Macmillan, London, second edn.
Hennock, E.P. (1973) *Fit and Proper Persons*, Edward Arnold, London.
Heraud, B.J. (1966) 'The new towns and London's housing problem', *Urban Studies*, Vol. 3, no. 1, pp. 8–21.
Higgins, J. (1978) *The Poverty Business, Britain and America*, Basil Blackwell, Oxford.
Higgins, J., Deakin, N., Edwards, J. and Wicks, M. (1983) *Government and Urban Poverty*, Basil Blackwell, Oxford.
HMSO (1977) *Policy for the Inner Cities*, London, Cmnd. 6845.
HMSO (1983) *Streamlining the Cities*, London, Cmnd. 9063.
HMSO (1985) *Lifting the Burden*, London, Cmnd. 957.
HMSO (1986a) HMSO (1985) *Lifting the Burden*, London, Cmnd. 9571. *Building Business . . . Not Barriers*, London, Cmnd. 9794.
HMSO (1986b) *The Conduct of Local Authority Business*, The Widdicombe Inquiry, London, Cmnd. 9797.
HMSO (1988a) *Action for Cities*, London.
HMSO (1988b) *The Conduct of Local Authority Business: the Government Response to the Report of the Widdicombe Committee of Inquiry*, London, Cmnd. 433.
Holman, R. (1971) 'The Urban Programme appraisal', *Race Today*, Vol. 3, pp. 227–9.
Holman, R. (1978) *Poverty: Explanations of Social Deprivation*, Martin Robertson, Oxford.

Holtermann, S. (1975) *Census Indicators of Urban Deprivation, Working Note 6*, Department of the Environment, HMSO, London.

Home Office (1970) *Community Development Project: Objectives and Strategy*, HMSO, London.

Home Office (1982) *Statistical Bulletin 20/82*, HMSO, London.

Home, R.K. (1982) *Inner City Regeneration*, E. & F.N. Spon, London.

Horne, M. (1988) 'Monitoring and evaluation of industrial promotion: experience at Telford', *Local Economy*, Vol. 2, no. 4, pp. 271–85.

House of Commons Environment Committee (1983) *The Problems of Management of Urban Renewal (Appraisal of the Recent Initiatives in Merseyside)*, Vol. 1, HMSO, London.

House of Commons Committee of Public Accounts (1986a) *Enterprise Zones*, HMSO, London.

House of Commons Committee of Public Accounts (1986b) *The Urban Programme*, HMSO, London.

House of Commons Employment Committee (1988) *The Employment Effects of Urban Development Corporations*, HMSO, London.

Howe, Sir G. (1978) 'Liberating free enterprise: a new experiment', Bow Group Meeting, Isle of Dogs, June.

Hudson, R. and Sadler, D. (1987) 'National policies and local economic initiatives: evaluating the effectiveness of UK coal and steel closure area re-industrialisation measures', *Local Economy*, Vol. 2, no. 2, pp. 107–14.

Huhne, C. (1988) 'When the poll tax drops on the mat it will be no less painful than the rates', *Guardian*, 27 April.

Islington Council (1987) *Taking the Initiative*, London.

Jackman, R. and Sellars, M. (1977) *The Distribution of the RSG; the Hows and Whys of the New Needs Formula; CES Review 1*, Centre for Environmental Studies, London.

Jackson, P.M. (1984) *Local Authority Capital Expenditure Controls*, Association of Metropolitan Authorities, London.

Jacobs, M. (1986) 'Community businesses: are their aims confused?', *Local Economy*, no. 2, pp. 29–34.

Johnson, D. (1988) 'An evaluation of the Urban Development Grant programme', *Local Economy*, Vol. 2, no. 4, pp. 251–70.

Joseph, Sir K. (1972) 'The cycle of deprivation', speech, Pre-School Association, 29 June. Reprinted in E. Butterworth and R. Holman (eds.) *Social Welfare in Modern Britain*, Fontana, London, pp. 387–93.

Jowitt, T. (1988) 'Towards Silicondale? Lessons from American experience for the Pennine towns and cities', in K. Dyson (ed.) *Local Authorities and New Technologies: The European Dimension*, Croom Helm, Beckenham, pp. 125–37.

JURUE (1986a) *An Evaluation of Industrial and Commercial Improvement Areas, Inner Cities Research Programme*, Department of the Environment, HMSO, London.

JURUE (1986b) *Evaluation of Environmental Projects Funded Under the Urban Programme, Inner Cities Research Programme*, Department of the Environment, HMSO, London.

Kellner, P. (1987) 'Labour's future – decline or democratic future?' *New Statesman*, Vol. 113, no. 2934, pp. 8–11.

Kennett, S. and Hall, P. (1981) 'The inner city in spatial perspective', in P. Hall (ed.) *The Inner City in Context*, Social Science Research Council/Heinemann, London, pp. 71–87.

Kerner, O. *et al.* (1968) *Report of the National Advisory Commission on Civil Disorders*, US Government Printing Office, Washington, DC.

King, D.S. (1987) 'The State, capital and urban change in Britain', in M.P. Smith and J.R. Feagin (eds.) *The Capitalist City*, Basil Blackwell, Oxford.

Klausner, D. (1987) 'Infrastructure investment and political ends; the case of London's Docklands', *Local Economy*, Vol, 1, no. 4, pp. 47–59.

Law, C. (1988) 'Public–private partnership in urban revitalisation in Britain', *Regional Studies*, Vol. 22, pp. 446–51.

Law, C.M. in association with Grime, E.K., Grundy, C.J., Senior, M.L. and Tuppen, J.N. (1988) *The Uncertain Future of the Urban Core*, Routledge, London.

Lawless, P. (1979) *Urban Deprivation and Government Initiative*, Faber & Faber, London.

Lawless, P. (1981a) *Britain's Inner Cities: Problems and Policies*, 1st edn, Paul Chapman Publishing, London.

Lawless, P. (1981b) 'The role of some central government agencies in urban economic development', *Regional Studies*, Vol. 15, pp. 1–18.

Lawless, P. (1986) *The Evolution of Spatial Policy*, Pion, London.

Lawless, P. (1988) 'Enterprise board: evolution and critique', *Planning Outlook*, Vol. 31, no. 1, pp. 13–18.

Layfield, F. (1976) *Local Government Finance*, HMSO, London, Cmnd. 6453.

Lea, J. and Young, J. (1982) 'Urban violence and political marginalisation: the riots in Britain, summer 1981', *Critical Social Policy*, Vol. 1, pp. 59–69.

Leaders of the London Boroughs of Greenwich, Lewisham, Newham, Southwark and Tower Hamlets (1979) *Local Democracy Works*, London.

Leclerc, R. and Draffan, D. (1984) 'The Glasgow Eastern Area Renewal Project', *Town Planning Review*, Vol. 55, pp. 335–51.

LEDIS (1986) *Trade Unions and Local Economic Initiatives*, Local Economic Development Information Service, *Overview B33*, The Planning Exchange, Glasgow.

Lever, B. (1982) *Industry and Employment in Urban Areas, Unit 10 D202, Urban Change and Conflict*, Open University Press, Milton Keynes.

Liverpool CVS (1978) *People in Partnership*, Liverpool Council for Voluntary Service, Liverpool.

London Borough of Newham (1987) *Memorandum of Agreement by the London Borough of Newham and the London Docklands Development Corporation*, London.

London Docklands Development Corporation (1987) *Corporate Plan*, London.

Lovering, J. (1988) 'The local economy and local economic strategies', *Policy and Politics*, Vol. 16, no. 3, pp. 145–58.

Lovering, J. and Boddy, M. (1988) 'The geography of military industry in Britain', *Area*, Vol. 20, no. 1, pp. 41–51.

MacLeary, A. and Lloyd, G. (1980) 'Enterprise zones; a step forward', *Estates Gazette*, Vol. 255, pp. 149–50.

MacLennan, D. (1987) 'Rehabilitating older housing', in D. Donnison and A. Middleton (eds.) *Regenerating the Inner City, Glasgow's Experience*, Routledge & Kegan Paul, London, pp. 117–34.

MacLennan, D., Munro, M. and Lamont, D. (1987) 'New owner occupied housing', in D. Donnison and A. Middleton (eds.) *Regenerating the Inner City, Glasgow's Experience*, Routledge & Kegan Paul, London, pp. 135–51.

McArthur, A.A. (1986) 'An unconventional approach to economic development', *Town Planning Review*, Vol. 57, p. 87–100.

McArthur, A.W. (1987) 'Jobs and incomes', in D. Donnison and A. Middleton (eds.) *Regenerating the Inner City, Glasgow's Experience*, Routledge & Kegan Paul, London, pp. 72–92.

McDonald, I. and Howick, C. (1982) 'Monitoring the enterprise zones', *Built Environment*, Vol, 7. no. 1, pp. 31–7.

McKay, D.H. and Cox, A.W. (1979) *The Politics of Urban Change*, Croom Helm, Beckenham.

Manchester City Council (1987) *Manchester Employment Plan: A Strategy for Employment*, Manchester.

Marsh, N. (1988) *Jobs in Our Cities*, National Council for Voluntary Organizations, London.

Mason, T., Spencer, K.M., Vielba, C.A. and Webster, B.A. (1977) *Tackling Urban Deprivation. The Contribution of Area Based Management*, Joint Centre for Regional, Urban and Local Government Studies, Birmingham.

Massey, D.B. (1982) 'Enterprise zones: a political issue', *International Journal of Urban and Regional Research*, Vol. 6, pp. 429–34.

Massey, D. (1984) *Spatial Divisions of Labour*, Macmillan, London.

Massey, D. (1987) 'Equal opportunities: the GLEB experience', in A. Cochrane (ed.) *Developing Local Economic Strategies*, Open University Press, Milton Keynes.

Massey, D.B. and Meegan, R.A. (1982) *The Anatomy of Job Loss*, Methuen, London.

Mawson, J. and Miller, D. (1986) 'Interventionist approaches in local employment and economic development: the experience of Labour local authorities', in V.A. Hausner (ed.) *Critical Issues in Urban Economic Development*, Vol. 1, Clarendon Press, Oxford, pp. 145–99.

Mellor, M., Stirling, J. and Hannah, J. (1986) 'Worker co-operatives: a dream with jagged edges?', *Local Economy*, no. 3, pp. 33–41.

Merseyside Development Corporation (1986) *Annual Report and Financial Statement*, Liverpool.

Metropolitan Borough of Knowsley (undated) *Economic and Employment Policy Review: Stage 1: Position Statement. Current Economic Problems and Council Initiatives*, Liverpool.

Meyer, P. (1986) 'Assessing improvement area policy', *Local Economy*, no. 1, pp. 35–43.

Midwinter, A. (1985) 'Five myths about local government spending', *The Planner*, Vol. 71, no. 2, pp. 69–71.

Mills, L. and Young, K. (1986) 'Local authorities and economic development: a preliminary analysis', in V.A. Hausner (ed.) *Critical Issues in Urban Economic Development*, Vol. 1, Clarendon Press, Oxford, pp. 89–145.

Milner Holland (1965) *Report of the Committee on Housing in Greater London*, HMSO, London, Cmnd. 2605.

Mishra, R. (1984) *The Welfare State in Crisis, Social Thought and Social Change*, Wheatsheaf, Brighton.

Monck, C.S.P. (1986) *Science Parks – Their Contribution to Economic Growth*, UK Science Park Association, London.

Monck, C.S.P., Porter, R.B., Quintas, P., Storey, D.J. with Wynarczyk, P. (1988) *Science Parks and the Growth of High Technology Firms*, Croom Helm, Beckenham, in association with Peat Marwick McLintock.

Moore, B.C., Rhodes, J. and Tyler, P. (1977) 'The impact of regional policy in the 1970s', *Centre for Environmental Studies Review*, no. 1, pp. 67–77.

Moore, C. and Booth, S. (1986) 'The Scottish Development Agency: market consensus, public planning and local enterprise', *Local Economy*, no. 3, pp. 7–19.

Moore, C. and Pierre, J. (1988) 'Partnership or privatisation? The political economy of local economic restructuring', *Policy and Politics*, Vol. 16, no. 3, pp. 169–78.

Moore, C. and Skinner, V. (1984) 'Community business – a new synthesis', *Public Administration Bulletin*, no. 46, pp. 54–70.

Morison, H. (1987) *The Regeneration of Local Economies*, Clarendon Press, Oxford.

Nabarro, R. (1978) 'Assessment of the Partnership programmes', in T.A. Broadbent (ed.) *Inner Area Partnerships and Programmes: The First Year's Experience*, Centre for

Environmental Studies, Policy Series 8, London, pp. 25–37.

Nabarro, R. (1980) 'Inner city Partnerships: an assessment of the first programmes', *Town Planning Review*, Vol. 51, pp. 25–38.

Nabarro, R. and McDonald, I. (1978) 'The Urban Programme: will it really help the inner city?', *The Planner*, Vol. 64, pp. 171–3.

National Audit Office (1986) *Report by the Controller and Auditor General, Department of the Environment, Scottish Office and Welsh Office, Enterprise Zones*, HMSO, London.

National Audit Office (1988a) *Department of the Environment, Derelict Land Grant*, HMSO, London.

National Audit Office (1988b) *Department of the Environment, Urban Development Corporations*, HMSO, London.

National Community Development Project (1974) *Inter Project Report*, CDP Information and Intelligence Unit, London.

National Community Development Project (1975) *Forward Planning*, CDP Information and Intelligence Unit, London.

National Community Development Project (1977) *The Costs of Industrial Change*, London.

NCVO *et al*. Working Party (1984) *Joint Action – The Way Forward: Community Involvement in Local Economic Development Report of a Joint Local Authority Association and National Council for Voluntary Organisation Working Party*, London.

Newcastle City Council (1987) *Newcastle/Gateshead, Inner City Partnership, Action Programme, 1987–90*, Newcastle upon Tyne.

Newcastle upon Tyne (1981) *Enterprise Zone Scheme*, Newcastle upon Tyne.

Nicholson, B., Brinkley, I. and Evans, A. (1981) 'The role of the inner city in the development of manufacturing industry', *Urban Studies*, Vol. 18, pp. 55–71.

Norcliffe, G.B. and Hoare, A.G. (1982) 'Enterprise zone policy for the inner city: a review and preliminary assessment', *Area*, Vol. 14, pp. 265–74.

O'Connell, C. (1986) 'The Stoke National Garden Festival 1986 – a catalyst for investment', *The Planner*, Vol. 72, no. 1, pp. 11–13.

O'Connor, J. (1973) *The Fiscal Crisis of the State*, St Martin's Press, New York, NY.

Office of Population Censuses and Surveys (1984) *Key Statistics for Urban Areas*, HMSO, London.

P.A. Consultants (1987) *An Evaluation of the Enterprise Zone Experiment*, Department of the Environment, Inner-Cities Directorate, HMSO, London.

Page, D. (1987) 'What an estate to get into?', *Roof*, Vol. 12, pp. 18–20.

Parker, J. (1985) 'The availability of government grants', *The Planner*, Vol. 71, no. 1, pp. 14–16.

Parkinson, M. and Duffy, J. (1984) 'The Minister for Merseyside and the Task Force', *Parliamentary Affairs*, Vol. 37, pp. 76–96.

Parkinson, M. and Wilks, S. (1986) 'The politics of inner city partnerships', in M. Goldsmith (ed.) *New Relations in Central-Local Research*, Gower, Aldershot, pp. 290–307.

Patten, J. (1987) 'Inner city areas', *Hansard Written Answers*, Vol. 112, no. 73, Col. 397.

Pearce, G. (1988) 'City grants – lessons from the Urban Development Grant Programme', *The Planner*, Vol. 74, no. 4, pp. 15–19.

Perry, M. and Chalkley, B. (1985) 'New small factories in the public sector', *Area*, Vol. 17, pp. 185–91.

Pinker, R. (1979) *The Idea of Welfare*, Heinemann, London.

Platt, S. and Lewis, J. (1988) 'Thatcher's blueprint for the inner cities', *New Society*, Vol. 83, no. 1315, pp. 23–4.

Plowden (1967) *Children and Their Primary Schools*, HMSO, London.
Policy Studies Institute (1985) *Police and People in London*, London.
Potter, S. (1987) 'British new town statistics 1985–87', *Town and Country Planning*, Vol. 56, no. 11, pp. 292–7.
Power, A. (1987) *The PEP Experience 2. The PEP Guide to Local Housing Management*, Priority Estates Project, Department of the Environment.
Raban, C. (1986) 'The Welfare State – from consensus to crisis?', in P. Lawless and C. Raban (eds.) *The Contemporary British City*, Paul Chapman Publishing, London, pp. 3–23.
Race and Class (1981) 'Rebellion and repression, notes and documents', Vol. XXIII, pp. 223–44.
Redfern, P. (1982) 'Profile of our cities', *Population Trends*, Vol. 30, pp. 21–32.
Rees, G. and Lambert, J. (1985) *Cities in Crisis*, Edward Arnold, London.
Rex, J. (1982) 'The 1981 urban riots in Britain', *International Journal of Urban and Regional Research*, Vol. 6, pp. 99–113.
Rhodes, R.A.W. (1980) 'Some myths in central-local relations', *Town Planning Review*, Vol. 51, pp. 270–85.
Ridley, N. (1987) 'No general extensions to enterprise zones', Department of the Environment, Press Release 536.
Roberts, P. and Noon, D. (1987) 'The role of industrial promotion and inward investment in the process of regional development', *Regional Studies*, Vol. 21, pp. 167–73.
Rogaly, J. (1976) 'Let the centres of cities wither away', *The Financial Times*, 12 October.
Royal Town Planning Institute (1976) *Planning and the Future*, London.
Royal Town Planning Institute (1979) *Planning Free Zones*, London.
Rumbold, A. (1986) 'Inner cities debate', *Hansard*, Vol. 97, no. 115.
Rutter, M. and Madge, N. (1976) *Cycles of Disadvantage. A Review of Research;*, Heinemann, London.
Saunders, P. (1981) *Social Theory and the Urban Question*, Hutchinson, London.
Saunders, P. (1982) *The State an Investor, Unit 25, D202; Urban Change and Conflict*, Open University Press, Milton Keynes.
Scarman, L. (1981) *The Brixton Disorders, 10–12 April 1981, Report of an Inquiry*, HMSO, London, Cmnd. 8427.
Scott, A.J. (1982) 'Locational patterns and dynamics of industrial activity in the modern metropolis', *Urban Studies*, Vol. 19, pp. 111–42.
Seebohm (1968) *Committee on Local Authority and Allied Personal Social Services*, HMSO, London, Cmnd. 3703.
Segal Quince Wicksteed (1988) *Encouraging Small Business Start-Up and Growth: Creating a Supportive Local Environment*, sponsored by the Department of Employment, HMSO, London.
Sellgren, J. (1987) 'Local economic development and local initiatives in the mid-1980s', *Local Government Studies*, Vol. 13, no. 6, pp. 51–68.
Sheffield City Council (1987) *Working it Out: An Outline Employment Plan*, Sheffield.
Shelter (1972) *SNAP. Another Chance for Cities*, London.
Shore, P. (1977) *Hansard*, 929, Col. 84, 5 May.
Shore, P. (1978) *Partnership in the City*, National Council of Social Service, London.
Sills, A., Taylor, G. and Golding, P. (1985) 'Learning by our mistakes: experimentation and monitoring in the inner area programme', *Local Government Studies*, Vol. 11, pp. 53–63.
Sinfield, A. (1973) 'Poverty rediscovered', in J. Cullingworth (ed.) *Problems of an Urban Society, Vol. III, Planning for Change*, George Allen & Unwin, Hemel Hempstead, pp. 128–41.

Solesbury, W. (1986) 'The dilemmas of inner city policy', *Public Administration*, Vol. 64, pp. 389–400.

Solesbury, W. (1987) 'Urban policy in the 1980s: the issues and arguments', *The Planner*, Vol. 73, pp. 18–22.

Spencer, K.M. (1980) 'The genesis of comprehensive community programmes', *Local Government Studies*, Vol. 6, no. 5, pp. 17–28.

Spencer, K.M. (1981) 'Comprehensive community programmes; the practical experience', *Local Government Studies*, Vol. 7, no. 3, pp. 31–49.

Spooner, S. (1980) *The Politics of Partnership, Planning Studies 6*, Planning Unit, School of Environment, Polytechnic of Central London.

Stewart, M. (1983) 'The inner area planning system', *Policy and Politics*, Vol. 11, pp. 203–14.

Stewart, M. (1987) 'Ten years of inner city policy, *Town Planning Review*, Vol. 58, no. 2, pp. 129–45.

Stewart, M. and Underwood, J. (1982) 'Inner cities: a multi-agency planning and implementation process', in P. Healey, G. McDougall and M.J. Thomas (eds.) *Planning Theory, Prospects for the 1980s*, Pergamon Press, Oxford, pp. 211–124.

Storey, D. (1981) 'New firm formation, employment change and the small firm: the case study of Cleveland County', *Urban Studies*, Vol. 18, pp. 335–45.

Storey, D.J. and Robinson, J.F.F. (1981) *Local Authorities and the Attraction of Industry: The case of Cleveland, Discussion Paper 29*, Centre for Urban and Regional Development Studies, University of Newcastle upon Tyne.

Sunderland Borough Council (undated) *Next Generation Industries: Investor's Guide*, Sunderland.

Sunman, H. (1987) *Science Parks and the Growth of Technology Based Enterprises*, UK Science Park Association, London.

Swansea City Council (1981) *Swansea Enterprise Park*, Swansea.

Swansea City Council (1986) *Swansea Enterprise Park, Monitoring Report 19*, Swansea.

Talbot, J. (1988) 'Enterprise zones: are there no lessons for inner city policy?' *The Planner*, Vol. 74, no. 2, pp. 64–7.

Taylor, S. (1981) 'The politics of enterprise zones', *Public Administration*, Vol. 59, pp. 421–39.

Teague, P. (1987) 'The European Social Fund and British labour market policies – the unfulfilled local authority dimension', *Local Economy*, Vol. 2, no. 1, pp. 3–13.

Teitz, M.B. (1987) 'Planning for local economic development', *Town Planning Review*, Vol. 58, no. 1, pp. 5–18.

Thomas, A. (1988) 'Measuring the success of worker co-operatives and co-operative support organisations', *Local Economy*, Vol. 2, no. 4, pp. 298–311.

Thrift, N. (1979) 'Unemployment in the inner city; urban problem or structural imperative? A review of British experience', in D.T. Herbert and R.J. Johnson (eds.) *Geography and the Urban Environment: Progress in Research and Applications*, Vol. 2, John Wiley & Sons, Chichester.

Topham, S. (1978) *Problems of the Re-Use of Industrial Buildings, Research Note 14*, Joint Centre for Regional, Urban and Local Government Studies, Birmingham.

Town and Country Planning Association (1979) *Inner Cities*, London.

Town and Country Planning Association (1986) *Whose Responsibility? Reclaiming the Inner Cities*, London.

Town and Country Planning Association (1987) *North-South Divide: A New Deal for Britain's Regions*, London.

Townsend, P. (1974) *The Cycle of Deprivation. The History of a Confused Thesis*, British Association of Social Workers, Birmingham.

Trades Union Congress (1988) *Trade Unions in the Cities*, London.

Trafford Metropolitan Borough (1988) *Enterprise Zones: Progress Reports*, Manchester.
Travers, T. (1985) 'A forest guide', *New Society*, Vol. 71, no. 1152, pp. 147–9.
Travers, T. (1987) 'Current spending', in M. Parkinson (ed.) *Reshaping Local Government*, Policy Journals, Hermitage, Berks., pp. 9–24.
Turok, I. (1988) 'The limits of financial assistance; an evaluation of local authortity aid to industry', *Local Economy*, Vol. 2, no. 4, pp. 286–97.
Tym, R. (1977) *Time for Industry: Evaluation of the Rochdale Improvement Area*, Roger Tym and Partners, HMSO, London.
Tym, R. and Partners (1984) *Monitoring Enterprise Zones: Three Year Report*, Department of the Environment.
Tym, R. and Partners (1987) *The Economy of the Isle of Dogs in 1987*, for Tower Hamlets London Borough, London.
Waldegrave, W. (1987) *Simplified Planning Zones*, Department of the Environment, press release 439, London.
Walker, P. (1979) 'A Conservative view', in M. Loney and M. Allen (eds.) *Crisis in the Inner City*, Macmillan, London, pp. 9–21.
Ward, R. (1987) 'London: the emerging docklands city', *Built Environment*, Vol. 12, pp. 117–27.
West Midlands County Council (1986) *Inner City Economic Policy: Case Study of the West Midlands, Research Paper no. 7.* Birmingham.
Whyatt, A. (1983) 'Developments in the structure and organisation of the co-operative movement: some policy considerations', *Regional Studies*, Vol. 17, pp. 273–6.
Widdicombe Inquiry (1986) *The Conduct of Local Authority Business*, HMSO, London, Cmnd. 9797.
Wildavsky, A. (1979) *The Art and Craft of Policy Analysis*, Macmillan, London.
Wolman, H. (1987) 'Urban economic performance: a comparative analysis', in V.A. Hausner (ed.) *Critical Issues in Urban Economic Development*, Vol. 11, Clarendon Press, Oxford, pp. 9–41.
Wood, P.A. (1986) 'The anatomy of job loss and job creation: some speculations on the role of the "producer service" sector', *Regional Studies*, Vol. 20, pp. 37–46.
Wray, I. (1987) 'The Merseyside Development Corporation, progress versus objectives', *Regional Studies*, Vol. 21, pp. 163–7.
Wright, M. and Coyne, J. (1985) *Management Buy-Outs*, Croom Helm, Beckenham.
Wright, M., Coyne, J. and Lockley, H. (1984) 'Regional aspects of management buy-outs: some evidence', *Regional Studies*, Vol. 18, pp. 428–31.
Young, Sir G. (1986a) *Simplified Urban Programme Proposals are Announced*, Department of the Environment, London, press release 437.
Young, K. (1986b) 'Economic development in Britain: a vacuum in central-local relations', *Environment and Planning C; Government and Policy*, Vol. 4, pp. 439–50.

Index